Sebastian Seitz

Magnetic Resonance Imaging on Patients with Implanted Cardiac Pacemakers

Vol. 10
Karlsruhe Transactions on Biomedical Engineering

Editor:

Karlsruhe Institute of Technology
Institute of Biomedical Engineering

Magnetic Resonance Imaging on Patients with Implanted Cardiac Pacemakers

by
Sebastian Seitz

Dissertation, Karlsruher Institut für Technologie
Fakultät für Elektrotechnik und Informationstechnik, 2010

Impressum

Karlsruher Institut für Technologie (KIT)
KIT Scientific Publishing
Straße am Forum 2
D-76131 Karlsruhe
www.ksp.kit.edu

KIT – Universität des Landes Baden-Württemberg und nationales
Forschungszentrum in der Helmholtz-Gemeinschaft

KIT Scientific Publishing 2011
Print on Demand

ISSN: 1864-5933
ISBN: 978-3-86644-610-6

Magnetic Resonance Imaging on Patientes with Implanted Cardiac Pacemakers

Zur Erlangung des akademischen Grades eines

DOKTOR-INGENIEURS

von der Fakultät für

Elektrotechnik und Informationstechnik

des Karlsruher Instituts für Technologie (KIT)

genehmigte

DISSERTATION

von

Dipl.-Ing. Sebastian Seitz

geboren in Marburg/Lahn

Tag der mündlichen Prüfung: 14.10.2010

Hauptreferent: Prof. Dr. rer. nat. Olaf Dössel
Korreferent: Prof. Dr. med. Evangelos Giannitsis

Abstract

Magnetic resonance imaging (MRI) has become an essential means of medical imaging in modern clinical diagnosis. Especially for the analysis of soft tissues it is often superior to other imaging techniques like computed tomography (CT) or X-ray. At the same time, the number of patients carrying a cardiac pacemaker or defibrillator is increasing. When these patients get in need for an MRI examination, their implanted medical device prohibits them now from being admitted to it. The reason for this contraindication are potentially dangerous interactions between RF fields occurring during MRI procedures and the leads and housings of the pacemakers. Besides dislocation and device malfunctions, which are both manageable today, the eventual heating of the tip of the leads is still regarded as an unresolved concern.

The aim of this work was to analyze and identify the patterns that might induce heating and to develop strategies to counteract or totally avoid them. Two approaches were taken: a computational one with simulations of the occurring electromagnetic field distributions and an experimental one with in-vitro studies using Plexiglas phantoms in real MRI devices. In compliance with previous results in the literature, a significant correlation between the position and length of the implants and the to be expected heating was found. Furthermore the computer simulations confirmed the in-vitro experiments and vice versa.

The most important finding of this work is the superiority of open MRI devices in contrast to the conventional bore-hole devices regarding a possible heating. The different orientation of the RF fields present in the open type reduced the induced currents in the pacemaker leads to a level where effects on the surrounding tissue are substantially less probable and hazardous.

Acknowledgment

Diese Arbeit ist zwischen Juni 2006 und Oktober 2010 am Institut für Biomedizinische Technik entstanden. Am Gelingen haben viele Menschen ihren Anteil, denen ich im Folgenden danken möchte.

Mein ganz besonderer Dank gilt meinem Betreuer, Herrn Prof. Dr. rer. nat. Olaf Dössel, für die Überlassung des Themas und die kontinuierliche, vertrauensvolle Unterstützung während der Durchführung dieser Arbeit. Unsere vielen spannenden fachlichen Diskussionen waren zugleich Herausforderung und Inspiration.

Ein weiterer herzlicher Dank geht an Herrn Prof. Dr. med. Evangelos Giannitsis für die Übernahme des Korreferats.

Danken möchte ich auch meinen Kolleginnen und Kollegen am Institut für das wunderbare und produktive Arbeitsklima sowie ihre Unterstützung in allen Verwaltungs-, Technik- und Literaturfragen.

SPEAG, Schmidt & Partner Engineering AG, gilt mein Dank für die Bereitstellung der verwendeten Software SEMCAD.

Am allermeisten möchte ich mich aber bei den lieben Menschen bedanken, durch die mir all dies erst ermöglicht wurde: bei meinen Eltern, meinen Großeltern und meiner Schwester mit ihrem Freund, die mich auf dem Weg hierher unermüdlich und unerschütterlich unterstützt haben und schließlich auch diese Ausarbeitung mit vielen kritischen Anmerkungen begleitet und bereichert haben. Und natürlich bei meiner Verlobten Joanna, dass sie mich immer wieder bestärkt und die Energie zu all diesem gegeben hat. Vielen Dank und dziękuję bardzo!

Karlsruhe, im Oktober 2010 *Sebastian Seitz*

Contents

1. Introduction

1.1. Motivation

Magnetic Resonance Imaging (MRI) is a widely used technique for medical imaging today. It offers unparalleled insights into soft body tissues. Due to the availability of static and motion pictures it is a valuable tool for the medical examiner to answer a wide spectrum of diagnostic questions.

At the same time, the improved life expectancy leads to a growing population of patients with cardiac diseases. During the course of their treatment, many receive an implanted cardiac pacemaker or defibrillator.

As shown later, numerous reports illustrate the potentially hazardous interaction between the electromagnetic fields present during MRI examinations and the system of electrode and pacemaker. The most dangerous is an excessive heating of the electrode tips. As a consequence, an implanted pacemaker till today means a clear contraindication for an MRI procedure.

This is an unfortunate situation because MRI is especially capable for diagnosing cardiac diseases. It offers a large portfolio of sequences specific to single pathologies like scars in the myocardium, ischemic tissue or perfusion dysfunctions in general after a myocardial infarction - which are often treated with a pacemaker. But this intervention till today blocks the use of MRI in follow-up examinations.

But the question of heating around metalic objects is not limited to already implanted pacemakers or defibrillators. Nowadays the navigation and orientation during the implantation of a pacing lead is mainly relying on the X-ray based C-arm. This is always connected with an exposure of the medical staff to a non-negligible amount scattered radiation. If a method or MRI protocol would be found, that allowed the use of MRI for interventional purposes, this would mean a significant advantage. Not only regarding the radiation, but also because MRI can provide three-dimensional images. The current position would not just be displayed in a flat projection discarding any spatial information.

1.2. Previous work at the Institute

The question whether patients with implanted pacemakers can safely undergo an MRI examination is by no means a new one. As described more detailed in chapter 2, initial work on this topic was for example done in 1981 by Davis et al. [1]. At our Institute of Biomedical Engineering (IBT), Karlsruhe, the first studies were carried out in 1999 by Golombeck [2]. One a of the findings achieved by computer simulation was the role of metal wires acting as short-cuts for induced voltages in the human body. Another one was a novel technique for determining current distributions caused by the gradient fields present during MRI. Eriksen et al. developed a computer model for a birdcage coil for high-field MRI applications using the numerical simulation software *MAFIA* by CST [3].

1.3. Aim of this work

The main aim of the presented work was to answer the questions directly deriving from the circumstances described above: How can patients with an implanted pacemaker still be enabled to safely undergo MRI procedures? What are the underlying mechanisms leading to heating at the tip of an electrode? How high are the occurring temperature elevations? What kind of modification in the scanning protocols, sequences or procedures can reduce heating?

Since the first studies at the IBT, the performance of computers has made a giant leap ahead. Hardware acceleration has speed up the calculation of electromagnetic fields in complex dielectric environments like anatomical voxel datasets significantly. Highly detailed models of pacemaker leads can be simulated and the effects of induced currents inside those models can be examined. Does the computation of those fine grained field distributions provide a substantial benefit in understanding the genesis of induced currents?

The development of realistic anatomical voxel models for numerical calculation of fields is a complex, time consuming procedure. Furthermore, it always provides only representations of one single individual. For dosimetry applications, the fast replication of all tiny details is often not necessary, an exact model of a subjects volume, that includes adapted organ models, would be more favorable. In this work, first steps were taken to develop a method to derive such models from a three-dimensional whole body scan.

1.4. Structure of this work

This work is structured in the following sections:

Chapter 1 presents the motivation and the main aspects of this work. It gives a short introduction into the topic and the work previously done at the IBT in this field.

Chapter 2 provides the status quo in literature on topics presented in this work. It covers research on interaction of implanted devices with the electromagnetic fields present during MRI, the exposure to radio frequency waves in general, the encapsulation of implants in non-excitable tissue, proposed changes to pacemakers and/or leads to reduce induced currents, the implications of using open MRI instead of the classic bore-hole type and general measurement issues.

Chapter 3 contains the theoretical background for all the methods used later. It describes the principles behind MRI and common excitation patterns, it illustrates the functionality of body coils that generate the rotating transversal B_1-field. Furthermore it introduces the underlying electromagnetic field theory and Finite Difference Time Domain (FDTD), the used numerical field calculation method. The differences between open and classic birdcage coils are illustrated, as well as the interaction of electromagnetic waves and biological tissues. A method for deriving temperature distributions from calculated electromagnetic (EM) fields is introduced, supported by basic information on the human thermoregulatory system and the frequency dependence of biological tissue. Finally, an overview is given on pacemaker types and commonly used pacing leads.

Chapter 4 covers the new methods developed in the course of this work. It is divided into computer simulations and in-vitro experiments. Here the used software SEMCAD is introduced as well as the phantoms providing the appropriate environment for the tested objects for later in-vitro and simulation studies. The development of the generic organ models is described, followed by the various CAD models of wires, pacemakers and leads. Possible modifications to the pacemaker-lead system are motivated. The chapter shows the newly developed open MRI coil and illustrates the idea, why open MRI could be superior to classic MRI in terms of induced currents. Finally a model for assessing the effects of fibrotic tissue around the leads is proposed and aspects and problems of evaluating the acquired results are highlighted. The second part of chapter 4 deals with the carried out in-vitro experiments. The environment of the tests it characterized as well as the employed methods, protocols, MRI types and measurement devices. The last sections of both parts deals with the implications of using open MRI devices instead of classic MRI to reduce the risk of induced currents.

Chapter 5 presents and discusses the generated results with the methods and procedures introduced in chapter 4. It is again separated into a part dealing with the simulations and a part on the in-vitro experiments. The computer simulations illustrate the validity of the developed open MRI coils model and the Huygens box approach. Here the results are discussed, computed for the detailed lead models, the encapsulated lead models and the pacemaker/lead model with lumped elements. The second part contains the measurement in the open and bore-hole MRI devices and describes the implications of the observed temperature distributions in both systems.

2. Status quo and Literature

The research on potential interaction of implants with the magnetic or electromagnetic fields present during an MRI procedure started soon after the introduction of the MRI to clinical practice [4]. Until today, the scientific community and also producers of all kinds of medical devices have spent an enormous effort to analyze how the MRI system can interact with objects present in or on the patients body. The question was not only if the occurring phenomena could cause any harm to the patient but also if they may lead to imaging artifacts rendering the acquired pictures unusable.

Today it is common sense that any kind of magnetizable objects in or on the body mean a clear contraindication for any MRI procedure and great care is taken to prevent them from getting close to the MRI device. This includes items like watches or prostheses and is also, for example, valid for patients with remains of shrapnels.

But besides these obvious aspects, the question was and still is whether patients with small implants like stents or active implants like pacemakers of cardiac defibrillators will be subject to hazardous effects, even if the implants consisted of non-magnetizable materials but were able to act as electric conductors.

In 1994, the American Society for Testing and Materials (ASTM) had proposed a standard test method for evaluating the effects of passive implants on MR images [5]. For small objects like dental implants made of titanium or gold, Klock et al. were able to show translational forces on orthodontic wires, that were 9.1–27.6 times higher than normal gravity [6]. Compared to the low weight of the wires, the effect was regarded as not hazardous. Regier et al. evaluated the heating near fixed orthodontic appliances in a 3 Tesla environment, but they only found miniature temperature changes of –0.3 to +0.2 °C degrees, which was seen as not dangerous at all for the patient [7]. One publication covering dental implants and their magnetic susceptibility in MRI nevertheless still recommended a per-case evaluation, because most but not all recent tested materials had been unaffected by the RF fields [8].

A second concern in recently published studies besides heating are imaging artifacts.

Due to more or less prominent disturbances of the imaging process by implanted objects, the aquired images could be rendered unusable [9, 10, 11, 12].

A good deal more vital is the research interest regarding active implants and risks for patients. The projects in this field can be grouped by the type of the magnetic respectively electromagnetic field that causes the potentially hazardous effects. During an MRI procedure, the patient is exposed to

- The static field

- Gradient fields

- RF fields

A detailed introduction to the generation and purposes of those fields will be given in section 3.1.

In general there are two approaches to this problem: to analyze the effects indirectly by monitoring the patient during and examining after an MRI procedure or directly by in-vitro studies and computer simulations. The first is more common for clinical and the latter for technical studies. The difference is caused by the fact that the region of interest – the tip of the electrode – is not accessible for fiber-optic measurements when the device has already been implanted. Therefore, only indirect effects like a change of stimulation thresholds, device malfunctions or an impairing of the patient can be studied.

The stimulus creating this wide research interest was and is that implants might cause local heating during an MRI examination. As described later, the integrity of human tissue is bound to a very narrow temperature range around $37\,^\circ$C. If the induced temperature increase significantly exceeds this level, the tissue is at risk for severe damage.

2.1. Static B_0-field

The most obvious field created by an MRI device is the static B_0-field. It is also officially addressed by an ASTM standard (F 2052) [13].

Two effects caused by this extremely strong field are in the first place a possible displacement of the device and/or the electrode and secondly the activation or dis-activation of the reed switch inside the pacemaker (see section 3.7 for more details on this topic).

In 1985, Dujovny et al. reported a deflection of up to 90° for aneurysm clips made of stainless steel and cobalt alloys [14]. More recently, Luechinger et al. tested force and torque effects on 31 pacemakers [15]. For devices released before 1995, they found magnetic forces of 0.05–3.6 N. For devices introduced later, no relevant forces were observed anymore. Two years later, Shellock et al. reported deflection angles of up to $90°$ for 14 different pacemaker models though [16]. Furthermore they described the effects as to be lower in long-bore MRI devices than in short-bore models but did not specify if this was only true for older or as well for recent device models.

In 2002 Luechinger et al. examined the influence on reed switches (see section 3.7 for more details) when exposed to the static field [17]. They found that the reed switch behavior of the tested pacemakers was difficult to determine when already enclosed in a pacemaker due to the unknown alignment of the switch relative to the B_0-field orientation.

2.2. Gradient fields and MRI-related ectopy

In 2005 Irnich et al. conducted an in-vitro study with pacemakers from deceased patients [18]. They found a strong influence of the magnetic fields on the reed switches. As a consequence, intentionally setting the pacemaker to a synchronous stimulation mode during the MRI procedure could avoid this potentially hazardous situation. Subsequently, 12 patients underwent MRI examinations without complications.

This work was referenced by Mollerus et al. when they administered a study with 52 patients with pacemakers who were subject to medically necessary MRI scans [19]. The authors identified only a vague relation between the MRI procedures and occurring ectopies but were unable to determine a reliable pattern.

In 2008, Tandri et al. presented a study where they only found very low currents caused in straightly lain out pacemaker leads. When the lead was coiled up (diameter approx. 2 cm), the induced currents increased to a level where myocardial capture could occur. The conclusions made by Tandri et al. were heavily criticized by Irnich [20], leaving the topic open for discussion.

The most recent work at the time of writing was reported by Bassen et al. in 2009 [21]. They developed an E-field probe and measured the induced fields near pacemaker leads in MR gradient fields. For abandoned leads with uncapped proximal endings and for a complete system with an isolation failure at the connection terminal, Bassen and co-workers described E-field distributions near the tips that were capable of unintended myocardial stimulation. Another effect was observed for gradient fields coinciding with a stimulation pulse from the pacemaker. In these cases the shapes of the pulses were significantly altered, both decreased and increased.

2.3. Radio-frequency related Effects

Shortly after the availability of MRI based on findings by Mansfield [22], Davis et al. carried out a compatibility tests for surgical clips and hip prosthesis when placed in saline and exposed to a 42 MHz EM-field [1]. They observed a temperature increase of less than 1.1 K for the small structures and 2.1 K–5.8 K for the prosthesis.

One of the first studies that examined the problem of heating around pacemakers systematically was presented in 1997 by Achenbach et al. [23]. 25 electrodes and 11 pacemakers were exposed in a 1.5 T device (Siemens Magnetom) to T1-weighted spin-echo sequences. The authors observed maximum temperature increases of up to 32.5 °C for an unconnected unipolar lead in free space This value decreased to 5.7 °C when connected to a pacemaker and further to 3.7 °C for when placed in 0.9 % NaCl solution. These findings were confirmed by Shellock and co-workers, who also found an increased temperature around unconnected leads as compared to leads with attached pacemakers.

In 1998, Ladd et al. reported about an in-vitro study with coaxial cables of different lengths (0–300 cm in air and 0–100 cm in NaCl solution, $\varepsilon_r = 80$) [24]. They identified a maximum heating for 140 cm in air and 45 cm in NaCl solution. The latter is equal to approx. $0.86 \cdot \lambda_{\mathrm{NaCl}}$ (see table 3.6 for a list of wavelengths in various media).

In 1999, Nyenhuis et al. described a combined in-vitro and simulation approach [25]. The simulations were computed a quasistatic solver, the authors regarded the size of the tested object (cylinder with a radius of 38 mm and a height of 28 mm) as too small compared to the wavelength at 64 MHz. It also included a reference to possible impacts of the skin effect but this aspect was not further investigated.

Still in 2008, Nordbeck and Bauer reported severe malfunctions in pacemakers occurring during MRI scans [26]. In their study with eight recent pacemaker types and seven ICDs (Implanted Cardioverter Defibrillator), four pacemakers signaled a request for device replacement because of dropped battery voltage and three an additional reset of programmed pacing parameters. The ICDs showed no errors under the same conditions. In the same year, Nordbeck and Bauer conducted a second study focusing on in-vitro measurements of E-field and temperature distributions in 1.5 T MRI environments [27]. The E-field was assessed with a custom made probe allowing arbitrary positions and orientations. The authors found that temperatures increased the most if the implants were aligned along the E-fields induced in the phantom. In contrast to Nordbeck's findings, Gimbel described the outcome of 16 MRI scans of patients with pacemakers in a 3 T system as being completely free of hazardous device related events [28]. One patient noted chest pain during a brain scan that resolved before the end of the procedure and did not cause any device malfunctions or persistent problems.

Stenschke et al. presented a numerical simulation study combined with experimental validation on hip prostheses. They also created computer models of the implant and compared the calculated temperature changes with in-vitro measurement. But instead of a birdcage coil, the team used a circular polarized wave based on two orthogonal plane waves for excitation. According to the authors, the achieved results matched the in-vitro experiments well [29].

A thorough in-vitro study on the effects of lead position, length and orientation was presented by Mattei et al. in 2008 [30]. By examining 374 different configurations, they identified patterns determining the induced currents in metallic implants. The highest values were found in wires placed the furthest from the center of the coil where the B-field showed the highest gradients over time. Furthermore, the length contributed significantly in the formation of induced currents and subsequent heating. To capture the temperature distributions, fiber-optic temperature probes were used.

The main parameter commonly used to determine and limit the energy deposited in a patient during MRI is the specific absorption rate (SAR, see section 3.3 for details). For regulatory purposes, the SAR values are commonly normalized with respect to 1 or 10 grams. When compared to the size of the regions where heating is initiated near metallic implants, a volume correlating with 1 or 10 grams can be very coarse. Mattei et al. examined the influence of the normalization volume and showed that the local SAR or a volume of 0.1 g could represent significant temperature changes more reliably [31].

The effects of various experimental aspects like phantom filling parameters, probe position and calibration were evaluated by Neufeld et al. in 2009 [32]. In this study, they quantitatively assessed those parameters and outlined the relative measurement error contributions of each parameter.

In 2010, Kolandaivelu et al. outlined an alternative use of MRI for temperature mapping [33]. Because the spin relaxation time contains a temperature dependency, the authors used it for the simultaneous evaluation of an RF ablation. The achieved results proved it to be a promising method. An in-depth review article on MRI based temperature mapping was published by MacDonald in 2005 [34].

2.4. Exposure to Radio Frequency Waves in the Megahertz Range

The majority of the research efforts on exposure to electromagnetic fields in the megahertz region is concentrated on the high end of this frequency range. The use of modern mobile communication devices stimulated a tremendous number of scientific projects but they focus on effects caused by signals between 850 MHz and 2400 MHz. In this case, the dielectric properties of tissue (see sections 3.6) and thus the response is significantly different. In consequence, the results provided only very limited support for the actual project.

The interaction of mobile phones and other RF emitting devices with implants has been examined by Irnich and co-workers [35] as well as Tandogan et al. [36, 37] or Seitz [38, 39]. These projects rather aimed at a possible malfunction or harm to the device and did not focus on heating caused by induced currents.

Adair and colleagues reported about a patient study in which volunteers were exposed to RF fields of 100 MHz [40]. No statistically significant heating was observed during the experiments. The same author presented an in-depth review article about thermoregulatory mechanisms [41]. Allen et al. presented a study design for parallel exposure of subjects and a FDTD simulation at 220 MHz that was later used by Adair et al. to conduct a study with 6 adult volunteers [42]. They could only find small changes of about $\leq 0.35\,^\circ\text{C}$ as response to an exposure with an averaged whole body SAR of 0.78 W/kg.

2.5. Inclusion of Electrodes in Non-excitable Tissue

When electrodes are implanted, after a while the surrounding tissue can start to encapsulate the tip and the feed line with a layer of fibrotic, non-excitable tissue. Although by increasing the stimulation threshold, this sheath could serve as a natural isolation preventing induced currents in the electrode from emitting into the myocardium. At the same time, the low conductivity could provoke an increased loss and subsequently heating in this layer. The genesis of fibrotic tissue is stimulated by a reduced blood perfusion caused by the insertion of a foreign body. This leads to a progressive degradation of the conductive properties of the affected region.

Grill et al. implanted two types of four-electrode array in six adult cats in the region of the loins and regularly measured the impedance [43]. One was based on epoxy, the other on silicone rubber. During the first 3 to 4 hours post implantation, the resistivity was as high as $1131 \pm 196\,\Omega$-cm $(0.09 \pm 0.51\,\text{S/m})$ for the silicone and $353 \pm 64\,\Omega$-cm $(0.28 \pm 1.56\,\text{S/m})$ for the epoxy array. In the following 1 to 4 days, it decreased to $269 \pm 29\,\Omega$-cm $(0.37 \pm 3.45\,\text{S/m})$ and $158 \pm 75\,\Omega$-cm $(0.63 \pm 1.33\,\text{S/m})$, respectively. Later it reached a rather constant level of $666 \pm 77\,\Omega$-cm $(0.15 \pm 1.30\,\text{S/m})$ for the silicone and $298 \pm 18\,\Omega$-cm $(0.36 \pm 5.56\,\text{S/m})$ for epoxy embedded array. All tested animals showed a very similar cellular response.

Stokes et al. examined the encapsulation process on 101 canines with implanted polyurethane leads (atrial and ventricular) [44]. They reported a significantly higher implant encapsulation in vessels with flow perturbation compared regions with to unobstructed blood flow. An even more inhomogeneous response of the surrounding tissue was described by Candinas et al. in 1999 [45]. Here, the inter-patient variability included electrodes that were nearly free of fibrotic tissue as well as extensively incorporated ones.

In total, data on dielectric properties of implant encapsulating tissue remains rare. Only Grill et al. reported values for frequencies up to 100 kHz, which is in the range of gradient fields but not B_1-fields.

2.6. Proposed design changes for pacemakers and electrodes

One of the first ideas modifying pacemaker leads in order to make them compatible with MRI was proposed by Jerzewski et al. in 1996 [46]: a replacement of the stainless steel shaft braiding with copper that was less magnetizable and had better conductivity properties. A disadvantage of copper wires could be that it is more brittle. In addition, the long term stability regarding constant bending stress needs further investigation.

Four years later Ladd et al. presented a new type of electrode with built-in coaxial chokes [47] as shown in figure 2.1a. They were modeled with respect to the wave-length of the RF field and caused a significant reduction of the occurring heating.

An approach that eliminates the problem of a wire in a time-variant magnetic field completely was proposed by Greatbatch et al. in 2002 (see figure 2.1b) [48]. They replaced the conventional metal based lead that normally captures the electromagnetic fields with a fiber-optic wire. To generate the necessary electric stimulus, a miniaturized power converter at the tip transformed the light energy back to electrical energy. A 150 mW gallium-arsenide laser was sufficient to provide the circuit with enough energy required for pacing. The power consumption of such a laser system is large and as a consequence would reduce the lifetime of the battery significantly.

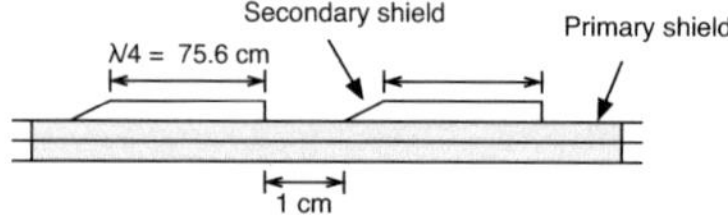

(a) Electrode with coaxial chokes as proposed by Ladd [47].

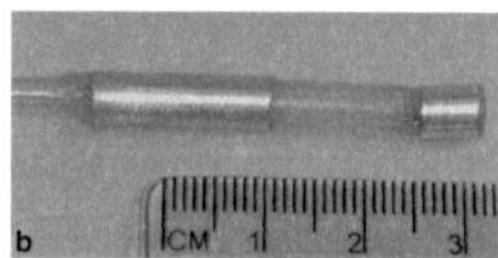

(b) Tip of fiber-optic pacing lead as proposed by Greatbatch [48].

Figure 2.1.: Modified and novel pacing lead designs.

Helfer et al. recommended a parallel circuit with a maximum impedance at the frequency of the transversal B-field (e. g. 64 MHz for a 1.5 T device) [49]. This could be achieved by either modifying the parameters inductance L, resistance R_s and capacity C_s of the lead or by adding discrete components (see figure 2.2).

One year later Stevenson [50] used a similar approach and developed a miniaturized bandstop circuit that could be integrated in the tip of a cardiac or neuro pacemaker (see figure 2.3). In the tested configuration, the so called *MRI Chip* reduced the occurring heating from 28.5 °C to 2.7 °C when compared to an unmodified pacemaker lead.

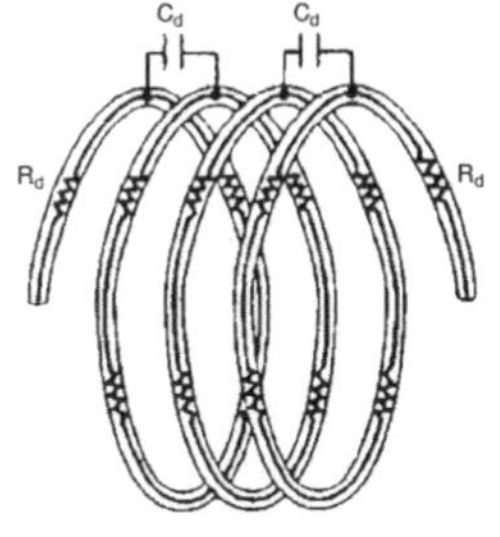

(a) Distributed capacitors and distributed resistors as present in a common lead wire.

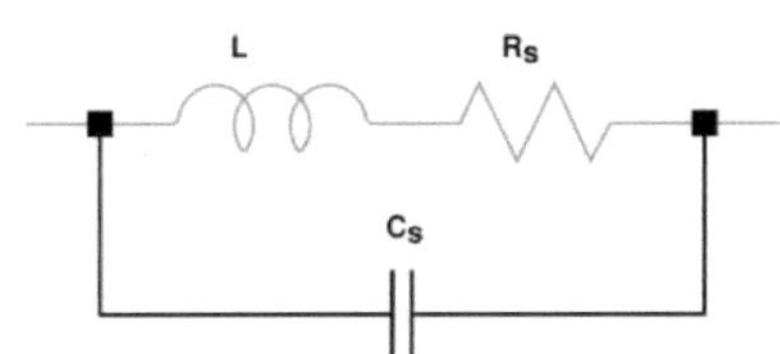

(b) Proposed equivalent circuit with parasitiv capacity, series inductor and series resistancy.

Figure 2.2.: Equivalent circuit of a pacemaker lead [49].

Tandri et al. proposed in a work mainly focusing on currents induced by low-frequency gradient fields the isolation of the pacemaker housing from the electronic circuit inside [51]. They observed a considerable reduction in occurring heating.

But not only within the scientific community, also in companies the problem is addressed. Pacemaker producers like Medtronic spend serious effort on developing products that are inherently safe regarding MRI. In 2009, Medtronic received approval for its pacemaker system *EnRhythm MRI* for head and extremity scans. In March 2010, they received approval in selected European countries for the *Advisa DR MRI* system for whole body scans without excluded regions.

This effort might at least partially be stimulated by the size and attractiveness of the target group. Due to the demographic development, more and more people are aspirants for implanted pacemakers and right now become unavailable for MRI examinations immediately at the moment of the implantation.

2.7. Open MRI

The research on the interactions between implants and open MRI devices is a relatively young field. In 2009, Luechinger et al. first examined such a configuration [52]. They observed significantly lower temperature changes compared to bore-hole devices. The

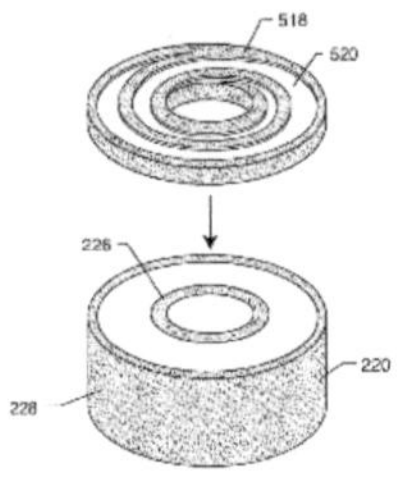

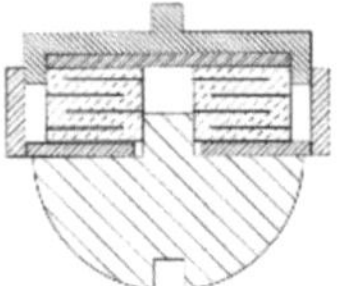

(a) Feedthrough capacitor with spiral inductor.

(b) *MRI Chip* integrated into passive fixation tip of pacemaker electrode.

Figure 2.3.: Pacing lead modifications as proposed by Stevenson (the numbers are part of the patent document)[50].

effects were more prominent in positions close to the boarder of the used phantom than in the center. The researchers attributed this to the changed orientation of the RF fields as also shown later in this work.

One year later, Strach et al. described a study on low-field examinations of 114 pacemaker patients [53]. The link to this work is, that open MRI devices also offer lower magnetic fields (1.0 T) compared to bore-hole models (>1.5 T). The study design aimed at examining patients before and after an MRI procedure as well as detecting arrhythmic episodes, changes in lead impedance or battery voltage. No critical effects were observed.

Although numerous works have addressed the problem heating, ectopies and device malfunctions, there is still no final consent on how cardiac and brain pacemaker can be modified to allow patients undergo MRI procedures. A large variety of effects have been observed, that can cause currents in the leads and subsequently lead to heating at the boundary of electrodes and surrounding tissue.

3. Fundamentals

3.1. Magnetic Resonance Imaging

Magnetic Resonance Imaging (MRI) is a medical imaging technique that is based on the alignment and relaxation of proton spins. Compared to Computed Tomography (CT) it does not expose the patient to ionizing radiation and provides a significantly better contrast for soft tissues. An example for an MRI image is shown in figure 3.1. The principle underlying MRI is commonly illustrated with an analogy to a rotating top. They both have an angular momentum (I for the proton and L for the top), but L has to be initiated mechanically while I is inherent to proton. In a space free of magnetic or electro-magnetic fields, the orientations of a magnetic moment μ of a nucleus are randomly distributed and all orientations are equal regarding energy. From a macroscopic perspective, they compensate each other. The magnetic moment is defined as

$$\vec{\mu} = \gamma \cdot \vec{I} \qquad [3.1]$$

with γ as gyromagnetic ratio. If the nucleus is placed in a homogeneous static magnetic field $\vec{B}_0$. It starts precessing when $\vec{B}_0$ and μ are not fully aligned. Additionally, the nucleus achieves an additional potential energy:

$$E = -\mu_z|\vec{B}_0| = -\mu_z B_0 \qquad [3.2]$$

This z-component of μ is quantized as

$$\mu_z = \gamma\hbar m \qquad [3.3]$$

with a nucleus specific range of m, e.g. $m = \left\{-\frac{1}{2}, +\frac{1}{2}\right\}$ for protons. As a consequence, equation 3.2 becomes

$$E = -\gamma\hbar m B_0 \qquad [3.4]$$

Because of the law of energy conservation, transitions between energy levels can only be induced by an external RF field. This interaction of the magnetic moment with an RF

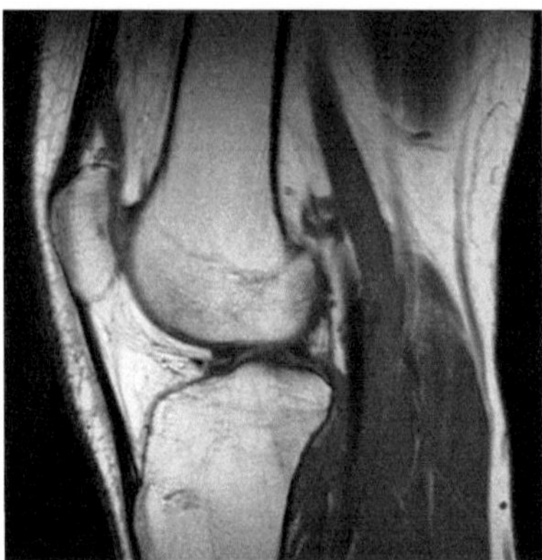

Figure 3.1.: MRI image of a human knee joint (Image: Wikimedia Commons).

field is called *Magnetic Resonance.*

The difference between two energy levels is $\hbar\omega_0 = \gamma\hbar B_0$, which results in the resonance condition [54]

$$\omega_{\text{RF}} = \omega_0 = 2\pi f_0 = \gamma B_0 \qquad [3.5]$$

The expression f_0 in this term is also known as *Lamor frequency*, a parameter specific for an isotope and the amplitude of B_0. For a hydrogen isotope in an 1 T field, f_0 is 42.577 MHz, at 1.5 T it is 63.86 MHz and at 3 T it is 127.6 MHz.

Since an orientation parallel to a present B field means a lower energy level, when exposed to the B_0 field in an MRI device, nuclei will align to that field. And when they get deflected by the perpendicular B_1 field later, the nuclei will return to that lowest energy. This process of re-alignment is described by two time constants (see figure 3.2):

- T_1 is the period of time in which the longitudinal magnetization returns to 63 % of its original equilibrium M_0 (parallel to B_0).

- T_2 is the period of time in which the transverse magnetization decays to 37 % of its original level.

The process that is characterized by T_1 is called spin-lattice relaxation and is strongly dependent on the Lamor frequency f_0. T_2 is caused by the mutual interaction of spins (spin-spin relaxation) and is nearly independent of f_0 (see table 3.1).

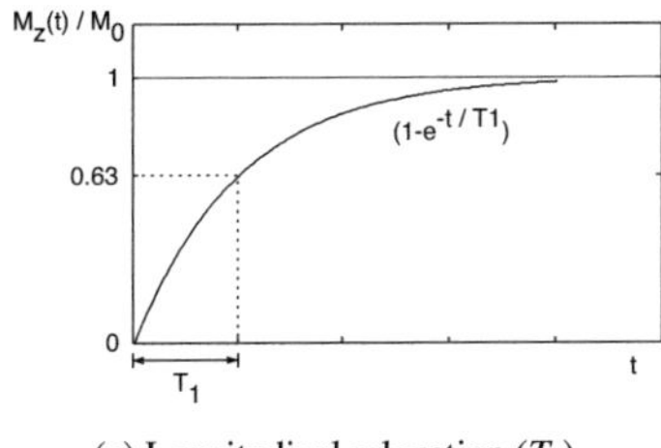

(a) Longitudinal relaxation (T_1).

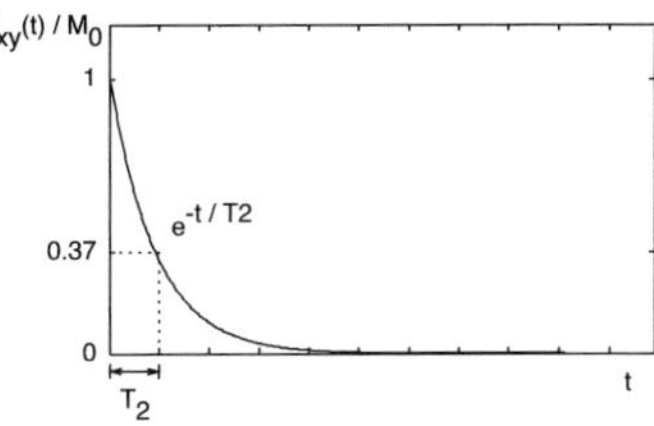

(b) Transversal relaxation (T_2).

Figure 3.2.: Relaxation of transversal and longitudinal magnetization.

Tissue	T_1 at 1.0 T	T_1 at 1.5 T	T_2
	[s]	[s]	[ms]
Skeletal muscle	0.73 ± 0.13	0.87 ± 0.16	47 ± 13
Heart muscle	0.75 ± 0.12	0.87 ± 0.14	57 ± 16
Fatty tissue	0.24 ± 0.07	0.26 ± 0.07	84 ± 36
White brain matter	0.68 ± 0.12	0.79 ± 0.13	92 ± 22

Table 3.1.: Relaxation times for selected biological tissues [54].

The deflection of the magnetization relative to the orientation of B_0 is called *flip angle*. It is adjusted by the time t_p in which the RF-field B_1 is switched on:

$$\alpha = \omega_1 t_p = \gamma B_1 t_p \qquad [3.6]$$

Common values for α are $90°$ and $180°$, the actual adjustment depends on the used pulse sequence. This parameter is of great importance for potential heating because it determines the activation times of the RF fields that may induce currents.

3.1.1. Electromagnetic Fields in MRI

In MRI, three different types of (electro-) magnetic fields are used:

- Static B_0-field

- Gradient B-fields (Kilohertz range)

- Orthogonal RF fields (e.g. 64 MHz at 1.5 T)

The B_0-field (see previous section for details) can be assumed as constant within the field of view (FOV), which is in the bore hole of the device (compare figure 3.3a). Only when the patient is moved in or out, the derivative of the static magnetic field dB_0/dt is unequal to zero and then could have an inductive influence on implanted devices. Further possible effect is the attraction of magnetizable elements like instruments, a dislocation of the pacemaker-lead system or the manipulation of the reed switch.

Because B_0 is homogeneous and time-invariant, it cannot contribute localization information to the signal. Therefore it is modulated with gradient fields in x, y and z direction. The z-gradient for example is described as

$$G_z = \frac{\partial B_z}{\partial z} \text{ or } B_z = B_{00} + G_z \cdot z \tag{3.7}$$

3.1.2. MRI Excitation Patterns

The time-series of RF bursts used for deflecting the spins follow a wide variety of medical questions. Depending on what is of special interest for the examiner, the excitation patterns or pulse sequences are defined. Imaging of a scar or fibrotic tissue requires different sequences than, for example, the visualization of blood perfusion. When compared to each other, the duty cycles differ significantly. As a consequence, the SAR is also varying. For patients with implants, a higher value for the on/off ratio means a prolonged period of possibly induced currents. In the following, a brief overview of common pulse sequences is to be given, a comprehensive can be found in [55].
The three *classical* sequences according to Dössel [56] and Reiser et al. are [54]:

- Saturation recovery

- Inversion recovery

- Spin-echo

3.1.3. MRI Coils

The RF fields used to deflect the spins are generated by coils arranged around the field of view and the region of interest. The coils can be devided in two different classes: receive/transmit coils and pure receive coils. The second ones are often flexible and therefor easier to apply on varying patient geometries. The outer shape depends on the intended type of use. There are coils especially designed for body, limb, head (neuro) and breast imaging.

Normally, fully integrated into every MRI device is a body coil. The simulations presented in this work also rely on this type, because it provides transversal RF fields that can be applied throughout the bore hole and covers large objects like the human torso and plexiglas phantoms. These coils work as receive and transmit coils. Besides the built in ones, body coils also exist as mobile units but in this configuration only receive signals.

The main purpose of the transmit coils is in all cases the generation of a rotating B-field whose vector is orthogonal to the B_0-field vector.

Bore hole MRI

In conventional bore hole MRI systems, the RF field required for flipping the spins is generated by a so called birdcage coil. The name is derived from the arrangement of the field emitting antennas in the shape of a birdcage as shown in figure 3.3b. The radio frequency signal is routed to the single antennas with a certain phase shift resulting in a rotating B-field. The frequency corresponds to the magnitude of the B_0-field as described in section 3.1.1. To ensure the resonating field inside the coil, also when loaded with a patient, a tuning of the coil with capacitors and resistances is necessary.

There are several publications available about how to design a birdcage. Ibrahim described how to design a birdcage head coil using FDTD [57]. In 2002, Eriksen et al. reported about a computer simulation study on dielectric resonators for applications at 7 and 12 T [3]. Still recently, a method was presented by the group of W. Zylka to develop MRI with assistance of numerical field calculation [58].

Open MRI

Since a few years, a second MRI device design is gaining attention: The open MRI. Because of the changed orientation of the static B_0-field, it has to employ a completely different technique to generate the transversal fields. The coils that provide B_1 are flipped by 90° and are now oriented horizontally in the plane of the patient tray (compare figure 3.4). In contrast to the birdcage, public knowledge about open MRI coil designs is rare.

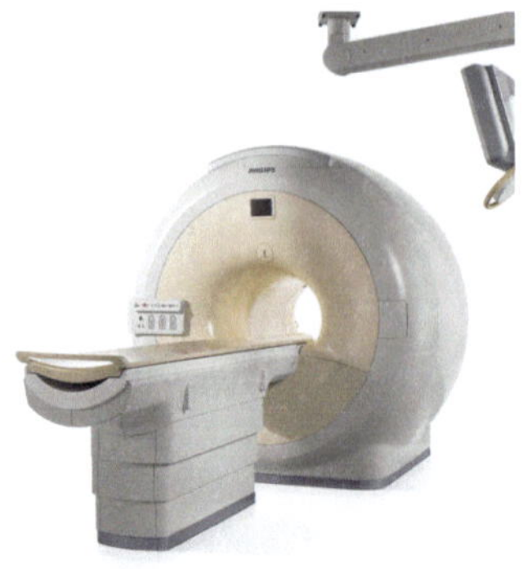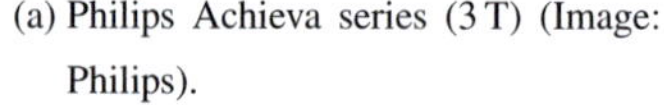

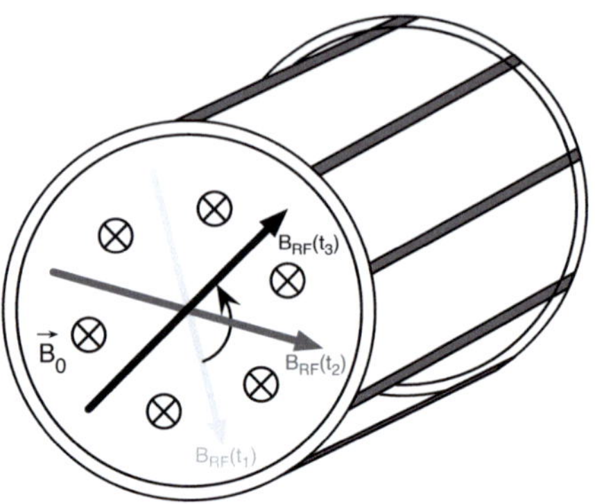

(a) Philips Achieva series (3 T) (Image: Philips).

(b) Schematic drawing of a birdcage coil inside a bore hole MRI device. The B_0-field is oriented along the axis of the coil and the HF B-field is rotating in the drawing.

Figure 3.3.: Classic bore-hole MRI system and schematic drawing of a birdcage coil.

Two publications deal with gradient coils [59, 60].

Fujita et al. proposed the use of a birdcage coil for open MRI, which actually collides with the intended free access from all horizontal directions [61]. McCarten reported about an open design for a body coil, but this aimed a MRI system with permanent magnets and does not comply with the field geometries in a real open MRI device [62]. Boskamp presented a coil design that contained two discs, combined with a network of capacitors, all together creating the rotating B_1 with a homogeneity better than 3 dB [63]. Parallel to this work, Khym et al. presented in 2010 a design comprising four rectangular stripes [64] near both – the upper and the lower – poles of the static magnetic field.

The boundary conditions for an open MRI coil are similar to the ones for a birdcage coil. It should generate a rotating field that is able to deflect the spins from the equilibrium state initiated by the B_0-field. Following the paradigm of the open construction, a coil design enclosing the field of view has to be avoided because it would collide with the patient tray. This limits possible constructions to the vicinity of the poles of the static magnetic field. The frequency of the B_1-field has to match the Lamor frequency associated with the B_0-field, in case of 1.0 T this would be 42 MHz.

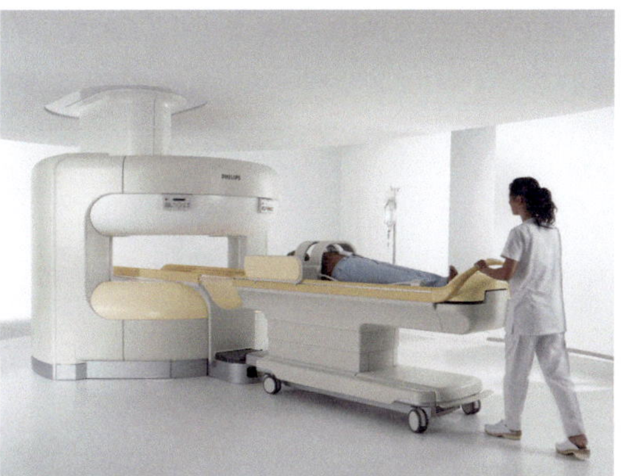

Figure 3.4.: Philips Panorama series (1 T) (Image: Philips).

3.2. Electromagnetic Field Theory

3.2.1. Maxwell's Equations

All computational approaches to determine electromagnetic field distributions ultimately have their basis in a set of differential equations. They were initially proposed by James C. Maxwell in 1865 [65] and cover the relation between electric and magnetic fields and their respective sources. The left sides shows the integral and the right side the differential form [66, 67, 68].

$$\oint_{\partial A} \vec{E} \cdot d\vec{s} = -\frac{\partial}{\partial t} \int_{A} \vec{B} \cdot d\vec{A} \quad \Leftrightarrow \quad \mathrm{rot}\,\vec{E} = \nabla \times \vec{E} = -\frac{\partial \vec{B}}{\partial t} \tag{3.8}$$

$$\oint_{\partial A} \vec{H} \cdot d\vec{s} = \int_{A} \left(\vec{J} + \frac{\partial \vec{D}}{\partial t} \right) \cdot d\vec{A} \quad \Leftrightarrow \quad \mathrm{rot}\,\vec{H} = \nabla \times \vec{H} = \vec{J} + \frac{\partial \vec{D}}{\partial t} \tag{3.9}$$

$$\oint_{\partial V} \vec{D} \cdot d\vec{A} = \int_{V} \rho\, d\vec{V} \quad \Leftrightarrow \quad \mathrm{div}\,\vec{D} = \nabla \cdot \vec{D} = \rho \tag{3.10}$$

$$\oint_{\partial V} \vec{B} \cdot d\vec{A} = 0 \quad \Leftrightarrow \quad \mathrm{div}\,\vec{B} = \nabla \cdot \vec{B} = 0 \tag{3.11}$$

In the special case of linear and isotropic materials the connection to material dependencies is achieved by the following terms

$$\vec{D} = \varepsilon\vec{E} \tag{3.12}$$

$$\vec{B} = \mu\vec{H} \tag{3.13}$$

$$\vec{J} = \sigma\vec{E} \tag{3.14}$$

While μ has a range of 0 (superconductors), 1 (vacuum) up to 500,000 (ferromagnetic materials), ε can vary between 1 (vacuum) and 100,000 (polymers) [69]. The values of σ are between 0 (vacuum) and infinity (superconductors).

The differential version of equation 3.9 provides the basis for an additional aspect. Applying the Nabla operator on both sides results in

$$\nabla(\nabla \times \vec{H}) = \nabla \cdot \vec{J} + \nabla \cdot \frac{\partial \vec{D}}{\partial t} \tag{3.15}$$

Because of $\nabla(\nabla \times \vec{H}) = 0$ and after switching spatial and temporal derivation, this is equal to

$$0 = \nabla \cdot \vec{J} + \frac{\partial}{\partial t}(\nabla \cdot \vec{D}) \tag{3.16}$$

Combining 3.10 and 3.16 gives

$$0 = \nabla \cdot \vec{J} + \frac{\partial \rho}{\partial t} \quad \text{or}$$
$$\nabla \cdot \vec{J} = -\frac{\partial \rho}{\partial t} \tag{3.17}$$

Expression 3.13 is also called the *electrical continuity equation* [70]. It is of importance when examining the effects of currents induced in wires that are placed in tissue or saline filled phantoms. In these cases, because no charge is deposited in the bare tips ($\partial\rho/\partial t$=0), all currents traveling along the wire will be emitted into the surrounding tissue.

Skin Effect and Proximity Effect

When a time-varying current is present in a conducting material, a phenomenon called *skin effect* can reduce the effective conducting area considerably. With increasing frequency, the currents are more and more compressed towards the outline of the wire

Materials	Resistivity	42 MHz	64 MHz	128 MHz
	[$\Omega \cdot$ m]	[μm]	[μm]	[μm]
Copper (Cu)	$1.68 \cdot 10^{-8}$	10.070	8.154	5.766
Gold (Au)	$2.44 \cdot 10^{-8}$	12.130	9.827	6.949
MP35N	$1.033 \cdot 10^{-6}$	78.931	63.942	45.214
Silver (Ag)	$1.59 \cdot 10^{-8}$	9.793	7.933	5.609

Table 3.2.: Skin depths for different materials [71].

reducing the current density in the interior area. The skin depth is defined as the distance between the surface of the conducting element and the point where the current density is reduced to $1/e$ of the current density at the surface. It can be determined with

$$\delta = \sqrt{\frac{2\rho}{2\pi f \mu_0 \mu_r}} \qquad [3.18]$$

with f as frequency, ρ as the resistivity in $\Omega \cdot$ m and μ_r as the relative permeability of the conducting material. It can be approximated as

$$\delta \approx 503 \sqrt{\frac{\rho}{f \mu_r}} \qquad [3.19]$$

A list of resulting skin depths is given in table 3.2. Besides the skin effect, the AC resistivity of a wire can increase even further if the wire is close to another conductor like in a coiled construction. The mechanism responsible for this is called *proximity effect*. In two proximate wires, the currents mutually induce eddy currents that influence the current distributions. They form and further suppress the conducting zone.

Since the RF fields applied in MRI use frequencies equal to or higher than 42 MHz, the skin effect is also of relevance for the presented work. The filigree wires inside the pacemaker leads are affected and their resistance increases remarkably. Furthermore, on the boundaries of tissue and wire, the current density could be concentrated and, as a consequence, be significantly higher compared to a completely pervaded cross-section.

3.2.2. Induction of Currents

The actual phenomenon that may lead to currents in metallic objects when exposed RF fields is the law of induction.

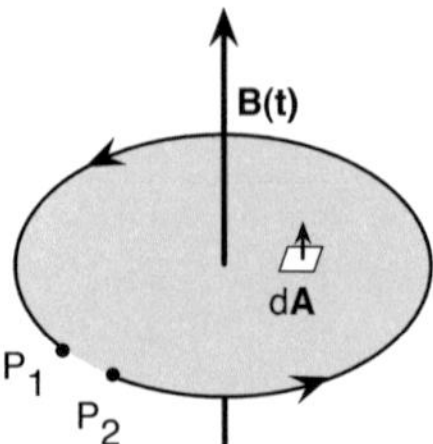

Figure 3.5.: Induction of voltage caused by time-varying magnetic field.

The law of induction, first proposed by Faraday, describes the fact that a current is induced in every conducting closed loop that is penetrated by a time-varying magnetic flux. It can be either provoked by moving the loop or by changing the intensity of the flux. This current can be quantified by splitting the loop and connecting a voltmeter. Mathematically induction is defined as

$$U_{ind} = -\frac{d\Phi}{dt} \qquad [3.20]$$

with Φ being the magnetic flux. The negative sign is caused by the law of energy conservation: the induced current is opposing its origin. This principle is also known as *Lenz' rule*.

The flux Φ can alternatively be expressed via the magnetic flux density $\vec{B}$ as

$$\Phi = \int \vec{B} \cdot d\vec{A}. \qquad [3.21]$$

The combination of both expressions yields to

$$U_{ind} = -\frac{d}{dt} \int_A \vec{B} \cdot d\vec{A} \qquad [3.22]$$

or for the configuration shown in figure 3.5

$$u_{P_1 P_2} = -\frac{d}{dt} \int_A \vec{B} \cdot d\vec{A} \qquad [3.23]$$

If the resistance R of the loop is > 0, equation 3.22 becomes

$$R \cdot i(t) = -\frac{d}{dt} \int_A \vec{B} \cdot d\vec{A} \qquad [3.24]$$

This principle is of very special interest for patients with pacemakers in MRI. When a lead, either with or without a connected pacemaker, is laid out in a loop inside the

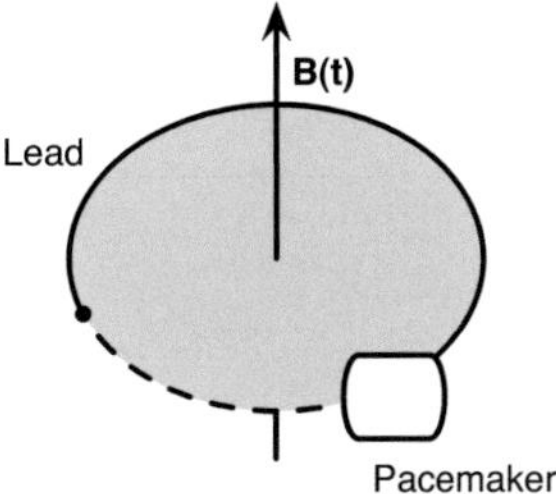

Figure 3.6.: Induction of voltage caused in pacemaker/lead system by time-varying magnetic field.

patients body and oriented perpendicular to the rotating B-field of the body coils, it is susceptible to induction. The loop is closed by the surrounding, even though just poor conducting, tissue 3.6.

As one can see in equation 3.24, the induced currents depend on the size of the loop formed by the lead, the resistance of the lead and pacemaker as well as the intensity of the time-varying magnetic field.

3.2.3. Comparison of Risk of Induced Currents in Open MRI and Birdcage Coils

One main question to be answered during this work was whether the RF field distributions in an open MRI coil led to less induced currents than the ones in a classic bore hole system. This idea was based on the fact that in an open MRI, the B_1-field vector is rotating in the same plane as the loop formed by the pacemaker-lead system is laid out (dotted line in figure 3.7). As a consequence, under ideal conditions there are no time-varying B-field components perpendicular to the loop and no currents will be induced. In a birdcage coil, the B_1-field vector's course is perpendicular to the pacemaker-lead plane (dashed line in figure 3.7) and therefore will generate a time-varying magnetic flux in the loop. In this case, currents may very well be induced in the lead.

If an open MRI system would be the solution for this problem it should therefore be evaluated with simulations and parallel in-vitro experiments in both kinds of MRI devices (see sections 4.1.11 and 4.2.7).

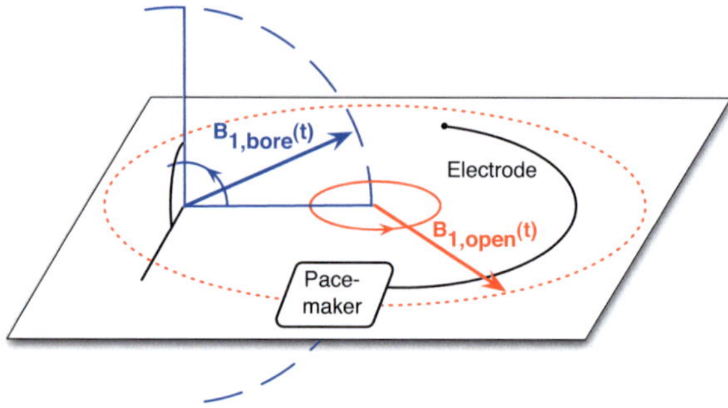

Figure 3.7.: Orientation of the pacemaker-lead system compared to the courses of the B_1-field vectors in an open (dotted line) and bore hole (dashed line) MRI.

3.2.4. Numerical Calculation of Fields

The calculation of electromagnetic fields with computers always requires the discretization of the problems in time and space. As a consequence, the continuous formulations of Maxwell's equations have to be transformed in a form that allows their solution not in an analytical but in a numerical way. A second aspect is the frequency range of the occurring EM waves. Can the problem be seen as static, quasi-static or time-variant? Many different approaches for this task exist, each with its specific advantages and disadvantages. The most popular as of today are:

- Finite Difference Time Domain (FDTD)

- Finite Integration Technique (FIT)

- Method of Moments (MoM)

- Finite Element Method (FEM)

As the MRI systems examined in this work operate with frequencies of 42 MHz respectively 64 MHz (see section 3.1.3 and 3.1.3), the analysis had to be carried out in the time-variant domain. In the course of this work, FDTD and FIT were used to compute EM field distributions.

Finite Difference Time Domain Method

The Finite Difference Time Domain method (FDTD) was presented by Yee in 1966 [72]. It implements Maxwell's curl equations in their differential form as a series of difference equations. In the following section, the method will be described for a Cartesian coordinate system, since computer implementation of the technique used in this work is also based on a Cartesian mesh.

In this case, equations 3.8 and 3.9 become

$$-\frac{\partial B_x}{\partial t} = \frac{\partial E_z}{\partial y} - \frac{\partial E_y}{\partial z} \qquad [3.25]$$

$$-\frac{\partial B_y}{\partial t} = \frac{\partial E_x}{\partial z} - \frac{\partial E_z}{\partial x} \qquad [3.26]$$

$$-\frac{\partial B_z}{\partial t} = \frac{\partial E_y}{\partial x} - \frac{\partial E_x}{\partial y} \qquad [3.27]$$

respectively

$$\frac{\partial D_x}{\partial t} = \frac{\partial H_z}{\partial y} - \frac{\partial H_y}{\partial z} - J_x \qquad [3.28]$$

$$\frac{\partial D_y}{\partial t} = \frac{\partial H_x}{\partial z} - \frac{\partial H_z}{\partial x} - J_y \qquad [3.29]$$

$$\frac{\partial D_z}{\partial t} = \frac{\partial H_y}{\partial x} - \frac{\partial H_x}{\partial y} - J_z \qquad [3.30]$$

For a numerical solution of these equations, the calculation domain has to be discretized in space and time [67]. For example B_z in 3.27 can be expressed as ,

$$B_z(x,y,z,t) := B_z(i\Delta x, j\Delta y, k\Delta z, n\Delta t). \qquad [3.31]$$

Thus 3.25 can also be defined as

$$\frac{B_x^{n+1/2}\left(i,j+\frac{1}{2},k+\frac{1}{2}\right) - B_x^{n-1/2}\left(i,j+\frac{1}{2},k+\frac{1}{2}\right)}{\Delta t}$$
$$= \frac{E_y^n\left(i,j+\frac{1}{2},k+1\right) - E_y^n\left(i,j+\frac{1}{2},k\right)}{\Delta z} - \frac{E_z^n\left(i,j+1,k+\frac{1}{2}\right) - E_z^n\left(i,j,k+\frac{1}{2}\right)}{\Delta y}$$

$$[3.32]$$

A second order finite-difference approximation was used for time and space. The "1/2" indexes lead to a mesh where the grids for E and H are shifted for half a mesh cell (see figure 3.8).

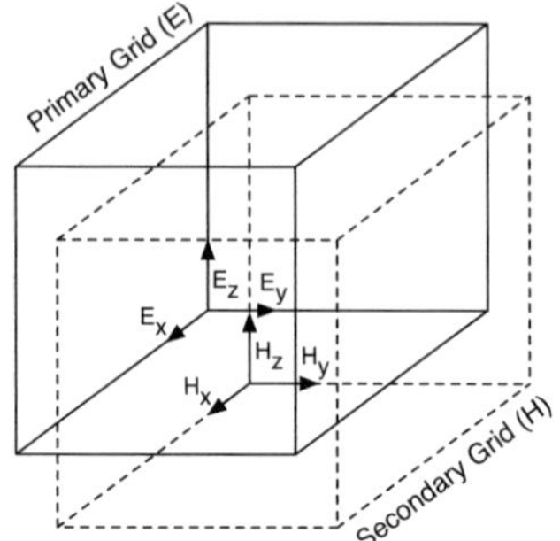

Figure 3.8.: Schematic view of FDTD grid as proposed by Yee [72].

The numerical stability criterion used in SEMCAD is the Courant-Friedrichs-Lewy condition (CFL). When using the staggered grid approach described above to solve Maxwell's equation, the CFL criterion becomes

$$\Delta t \leq \frac{1}{c\sqrt{\frac{1}{(\Delta x)^2} + \frac{1}{(\Delta y)^2} + \frac{1}{(\Delta z)^2}}} \qquad [3.33]$$

Here, c is the speed of light in the medium and Δx, Δy, Δz the outer dimensions of the smallest mesh cell found anywhere in the mesh. As one can see, a reduction of the mesh step size on the right side of unequation 3.33 has a direct impact on the constraints of the time-step.

The vectors $\vec{B}$, $\vec{D}$ and $\vec{J}$ all contain material specific factors (see equations 3.12–3.14) and therefore each of the mesh cells has to be attributed with values for conductivity and permittivity. For cubic structures perfectly aligned and matching the grid, this is no problem, but is most cases the objects of interest do not perfectly comply with the mesh. As a consequence, further steps are necessary to appropriately represent real world objects in a discrete mesh. In figure 3.9, several approaches are illustrated. When setting up a simulation, a fair balance between the level of geometric adaption especially of curved structures and number of mesh cells and therefor computation time has to be maintained.

The distances between grid lines do not have to be equidistant throughout the whole calculation domain as illustrated in figure 4.2, but FDTD does not allow a nested region with finer grid completely surrounded by a coarser one.

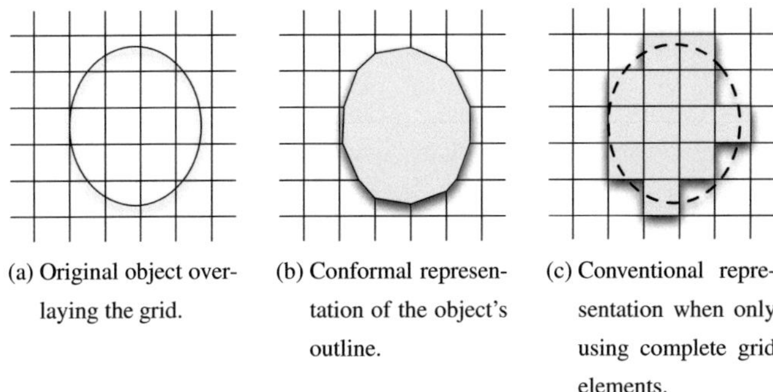

(a) Original object over-
laying the grid.

(b) Conformal represen-
tation of the object's
outline.

(c) Conventional repre-
sentation when only
using complete grid
elements.

Figure 3.9.: Grid representations of an elliptic object using different mesh generation methods.

The volume where all objects got placed and the software later computed EM field distributions, is called *calculation domain* or *computational domain*.

Boundaries

The volume of the calculation domain has to be finite to make it accessible to discretization and computations. Because of its cuboid shape, there are six boundaries that can interact with incident EM waves. In this work, each boundary was configured to completely absorb all incident fields. This can be achieved with a technique called *Perfectly Match Layer* (PML). Here, a set of absorbing layers rapidly attenuates incoming waves without reflection and by this emulates a calculation domain with infinite extent [68].
An example for other types of boundary models are *Perfectly Conducting Boundaries* with zero tangential E-fields. One usage scenario could be the guidance of plane waves. In all simulations presented later, only PML boundaries were implemented.

Sources

The point of origin for all EM wave propagations in a calculation domain are the sources where the exciting signals can be injected. Three major types of sources commonly exist in numerical field calculation applications like SEMCAD:

- Edge sources

- Plane wave sources

- Waveguide sources

The only ones of interest for this work are edge sources. In this case, an electric field strength is fed into an edge of the Yee grid [68]. The behaviour of the source can be either configured via current or via voltage. In the second case, the source also includes an internal resistance that absorbs energy out of the grid that is scattered back onto the source.

Total-Field / Scattered-Field

When setting up a simulation that will be computed using FDTD, the user faces the question how fine to adjust the mesh to properly represent objects contained in the calculation domain. If the project contains very small and at the same time very large objects, the small objects can lead to a simulation size and complexity impossible to compute on recent hardware. Original FDTD implementations provided no sufficient solution for this problem leading to a situation, where only coarse field distribution could be determined although miniature structures actually should to be analyzed. More recently, the *Total-Field / Scattered-Field* (TF/SF) approach offers a solution for this problem. The idea of the TF/SF method is to split the calculation into two steps. The first simulation, with the fine object not yet present, can be set up with a relatively coarse mesh (see figure 3.10b). Around the later position of the filigree object, the field distribution is determined and recorded, for example with a field monitor. In the second step, the calculation domain is restricted to a section around the small object (see figure 3.10c). The just recorded field distribution serves as a boundary condition and acts as a field source. In SEMCAD this type of source is called Huygens box. Due to the significantly smaller dimension of the second simulation, the filigree objects can be now meshed appropriately with a finer grid.

Important to consider when adjusting the field monitor is that the EM fields induced by objects inside the *Huygens box* must not touch the boundaries. If objects scatter the field towards of the boundaries and beyond, the boundary conditions of the Huygens Box would be infringed.

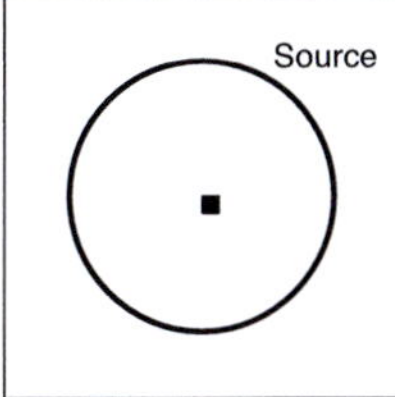

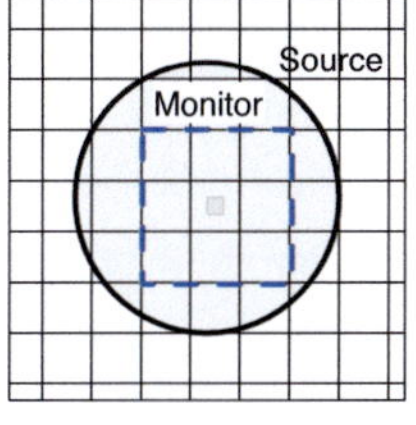

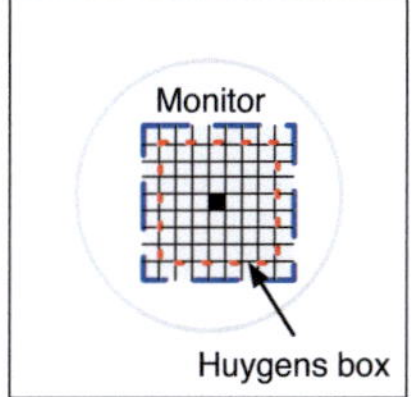

(a) Small object inside large circular source.

(b) Coarse mesh with active circular source and field monitor around disabled small object.

(c) Active small Huygens box inside disabled large circular source with fine mesh. Recorded field provides input for Huygens box.

Figure 3.10.: Total-Field / Scattered-Field as implemented by SEMCAD Huygens Box.

3.2.5. Hardware Acceleration

The solving of very large sets of linear equations and partial differential equations (PDE) as used in FDTD is expensive in terms of time and computational cost. One approach to handle this problem is described in section 3.2.4 and aims at reducing the complexity of the simulation. But at one point, the geometrical aspects of a configuration set a limit on how coarse a mesh can be while still representing the objects of interest appropriately. In this case only additional computing resources can reduce the simulation times. Fortunately, solving PDEs is a numerical problem that can be parallelized. The calculation is divided into smaller units that are distributed to a cluster of processes. Each process computes its share and returns the result back to the controlling system. The parallelization always brings up the question of communication overhead: does the benefit gained by processing many parts at the same time still prevail the now required synchronization?

The utilized software SEMCAD offers two options for parallelization: the use of a cluster of regular workstations or the deployment of the calculations to specialized hardware. For SEMCAD, the first technique scales nearly directly with the number of additional nodes, provided that all nodes share the same the architecture. So if a regular workstation (2x Intel Xeon 5150 @ 2.66 GHz, 24 GB RAM) can process approx. 40 MCells per second, two will achieve a performance of 80 MCells/s and so on.

Besides the cluster of workstations, nowadays another technology is becoming more and more popular for solving numerical problems. When visualizing or rendering a three-dimensional scene for a computer game or any other software, the scene can be partitioned and every part rendered independently. Due to the increasing popularity of computer games, the development of graphics processing units (GPU) used for this tasks has left behind that of general purpose central processing units (CPU). The instruction set of GPUs is drastically streamlined compared to regular CPUs but the number of processing units of recent models reaches up to 512 in one single chip.

The system employed in this work was built by Acceleware and relied on the *CUDA* framework of the GPU producer NVidia.

3.3. Specific Absorption Rate

When traveling electromagnetic fields collide with an object of different dielectric properties as the recent one, a part of the waves is reflected and the rest enters the object. If the penetrated object consists of a lossy material regarding EM propagation, a part of the field's energy is deposited in the object as thermal energy or heating.

The dimension of the SAR is *Watt per Kilogram* which gives a direct reference to the purpose of this measure: the quantification of absorbed energy in a certain portion of tissue. In the European Union, it is averaged over 10 g and in the United States over 1 g. In biological tissue, this absorption is commonly measured as *Specific Absorption Rate* (SAR) and can be determined in three different ways. Numerically the SAR is either calculated from the E-field vector $\vec{E}$ or the current density $\vec{J}$.

$$\text{SAR} = \frac{1}{2}\frac{\sigma|\vec{E}|^2}{\rho} \qquad [3.34]$$

$$\text{SAR} = \frac{1}{2}\frac{|\vec{J}|^2}{\rho\sigma} \qquad [3.35]$$

Here, σ is the electric conductivity and ρ the density of the target area. Experimentally, the SAR can be derived via the observed temperature change with c_i being the heat capacity of the object.

$$\text{SAR} = c_i\frac{dT}{dt} \qquad [3.36]$$

The last described method is only suitable for longer lasting heat incorporation but short compared to heat distribution processes because it includes a kind of averaging that obliterates short peaks of energy deposition.

The SAR is used as a measure in guidelines and regulatory standards to specify for the exposure of people in EM fields. The standards differentiate between whole body and local, occasional and occupational exposure. For the occupational case, there are different limits compared to therapeutic and general applications. Neuralgic areas like head require lower limits than for example arms, because in the first case too much deposited energy and as a consequence heating could cause damage easier. In research focusing on exposure to mobile phones, regions like eye, brain or ear therefore receive special interest [73, 74, 75, 76]. The reason of the observed effects is the limited blood perfusion and following that the risk of a local accumulation of the absorbed energy. Besides the use in MRI applications, the SAR is also employed, for example, for the regulation of RF emitting devices like mobile phones to protect the user from hazardous exposure situations. Special regulation for MRI examinations can be found in IEC/DIN 60601-2-33 [77], the limits for mobile devices have been defined in IEC/DIN 62209-1[78]. Some selected values have been listed in table 3.3. The SAR values are not only averaged over a specific tissue mass but also over time, the values in table 3.3 all include an averaging over 6 minutes. This may softening down high peaks in energy deposition only present for a few seconds. Although they can lead to serious tissue damage in the vicinity of implants when inducing currents, their occurrence is blurred by the averaging.

3.4. Thermoregulatory System

The human being as a homoiothermal life form has a highly developed system for regulating its body temperature. The main aspiration of this thermoregulatory system is to maintain the body's core temperature within a very specific narrow range. For healthy subjects, this is around $37\,°C$ with changes of $\pm0.6\,°C$ during the day [79]. At the same time, the temperature of the limbs can vary much more due to the surrounding temperature without harm for the organism. All this effort is done because tissue cells and the metabolic processes in them react very sensitively on heat. Should the temperature in the

General body exposure (100 kHz–10 GHz)			
Whole body	0.08		
Local SAR (head and trunk)	2		
Local SAR (Extremities	4		
MRI			
Body region	Whole body SAR	Partial body SAR	Local SAR (trunk)
Normal	2	2-10	10
1st level controlled	4	4-10	10
2nd level controlled	> 4	> (4-10)	> 10

Table 3.3.: Limits of specific absorption rate (SAR) for MRI applications. All values are in W/kg and averaged over 10 g [77, 78].

cell exceed 42 °C, it could initiate a process called apoptosis. This is an auto-destruction mechanism finally leading to the the cell's own death. When heated even up further, from 45 °C on the cells become subject to necrosis, the premature death of the cell [80]. A profound literature overview on the various aspects and thresholds of hyperthermia has been given by Roth [81].

Thus it is of special interest for this work, how the thermoregulatory system of the human body handles additional incorporated thermal energy like heating around the tip of the lead or a pacemaker housing during MRI.

Without moving particles, heat is dissipated by conduction. For a porcine heart, the thermal conductivity is between 0.6 and 0.75 W/m·K [82] which coincides with the value for water (0.6 W/m·K). This is the most import transport mechanism, when it comes to the pacemaker leads with a helix fixation mechanism. Because they are buried in the endocardium, they are separated from the massive blood flow in the inner volume of the heart. Thermo simulations without blood flow or convection solely rely on this heat transport mechanism. Because it is the least effective one, those simulations will always provide a worst case scenario.

The next higher level is convection. This describes heat transfer based on moving particles. In the human body, this is achieved by blood flow. The heart produces a cardiac output of ca. 5.6 l/min (*heart rate* × *stroke volume*, $70\,\text{min}^{-1} \times 0.081$) [79] and by doing this a rapid distribution of incorporated thermal energy away from its origin is guaranteed. Hence heating occurring in the inner heart is less hazardous than temperature rises in the myocardium.

3.5. Pennes Bioheat Equation

After determining the electromagnetic field distributions inside the body, the next step
is to derive a quantification for the heating induced by the deposed field energy. The
technique used in SEMCAD is based on the work of Pennes. In 1948, he proposed
a formulation how one can combine heat transfer, metabolism, perfusion and external
heat source (in this case the electromagnetic field) to calculate the resulting temperature
behaviour [83].

$$\rho c \frac{\partial T}{\partial t} = \nabla \cdot (k \nabla T) + \rho Q + \rho S - \rho_b c_b \rho \omega (T - T_b) \qquad [3.37]$$

Here S is the specific absorption rate, ρ the density, c the specific heat capacity, k the
thermal conductivity, ω the perfusion rate and Q the metabolic heat generation rate [68].
The first part contains an expression also known as *Fourier's law*. It establishes a direct
relation between the heat flux $\vec{q}$ and the negative temperature gradient ∇T.

$$\vec{q} = -k \nabla T \qquad [3.38]$$

The heat generated by metabolic processes (ρQ) during the time span of a MRI examination is small compared to the heat energy deposited by the RF fields and therefor can
be omitted.

The last part of equation 3.37 describes the blood flow in the volume of interest. Regarding MRI induced heating around implants this is a very crucial aspect, because the
higher the perfusion, the better the heat dissipation. Unfortunately little is known about
how to reliably quantify this blood flow. It depends on multiple parameters like location
(atrium or ventricle) or moment of contraction cycle. Furthermore, the electrode's type
of fixation can have a significant influence. When the tip is buried in the myocardium,
the perfusion is drastically reduced compared to an electrode fixed superficially with
hooks that is constantly washed around with new blood. SEMCAD includes a tool to
model a vessel system that can be parametrized with arbitrary blood perfusion rates –
due to the lack of information, this was not incorporated in the temperature simulations.
Nevertheless, any amount of perfusion reduces occurring temperature elevations. As a

consequence, crossing out this term will lead to a worst-case scenario. If the observed heating even then would not exceed thresholds for tissue damage, any real-life scenario could be assumed as reasonably safe.

Taking all those assumptions into account and combining it with equation 3.34, expression 3.37 becomes

$$\rho c \frac{\partial T}{\partial t} = \nabla \cdot (k \nabla T) + \rho S = \nabla \cdot (k \nabla T) + \frac{1}{2} \sigma |\vec{E}|^2 \qquad [3.39]$$

It establishes a direct connection between the RF fields and the occurring temperature change. As it still contains a time derivative on the left side, it forms a diffusion equation. When a time constant or steady state is reached ($\partial t = 0$), term 3.39 becomes a Poisson equation.

The potential risk associated with energy deposited in tissue is commonly assessed with the Arrhenius equation. It was originally developed by Arrhenius to evaluate the dependence of rate constants on the absolute temperature in chemical formulations. The original term for the rate constant k is given by

$$k = A \cdot \exp\left(-\frac{E_A}{RT}\right) \qquad [3.40]$$

Here E_A is the material specific energy barrier that has to be exceeded with a frequency of A, R is the universal gas constant and T the absolute temperature. This can be adapted to determine the accumulated damage caused to tissue over all times

$$\Omega(t) = A \int_{-\infty}^{t} \exp\left(-\frac{E_A}{RT(\tau)}\right) d\tau \qquad [3.41]$$

Possible applications are impacts of laser-light [84, 85, 86], RF ablation [87, 88, 89], electroporation [90], electrical trauma [91] or heat in general [92] on tissue.

3.6. Dielectric and Thermal Properties of Tissue and Other Used Materials

The calculation of field distributions in biological tissue requires detailed models of the dielectric properties for each occurring tissue type. When using computer models of anatomical structures as described in section 4.1.3, every voxel is linked to a set of values for conductivity and permittivity at the frequency of interest. The latter is important because conductivity and permittivity can vary significantly with frequency.

The most commonly used model was proposed by Gabriel et al. in 1996 [93, 94, 95]. It includes a mathematical formulation to calculate conductivity and permittivity for various tissue types at arbitrary frequencies. In this work, an application was developed that implemented this model. Table 3.4 lists computed values for selected tissues at 42, 64 and for comparison at 900 and 1800 MHz used by cell phones.

Besides human tissue, several other materials were used in the presented studies. The models described in section 4.1.3 and 4.1.6 contained plexiglas, isolation materials like silicone rubber and different metals. Furthermore the models were placed in saline solutions. The properties of those materials are listed in table 3.5.

3.7. Pacemakers - Function and Types

The motivation for implanting a pacemaker into a patient is to reestablish or maintain a regular rhythm in the excitation of the heart. Various diseases can lead to an irregular or too slow auto-stimulation. The pacemaker imitates the principle of the heart's own excitation system and emits short current pulses to initiate action potentials in the myocardium. There are different ways of how the current is actually applied in the patient:

- Transcutaneous with external electrodes on the chest

- Via electrode in the ocsophagus

- Temporal intracardial stimulation

- Permanent intracardial stimulation

In this work, the focus was placed on the last one: the stimulation with permanently implanted devices. All the others could possibly be removed during an MRI procedure and were therefore of reduced interest.

A pacemaker system consists of the main unit including battery, micro-controller, voltage amplifier and one or more connected electrodes. The number of electrodes depends on the required stimulation pattern. A detailed overview of electrodes is given in section 3.8. The pacemaker is normally packaged in a titanium housing to ensure biocompatibility. The outlets for the electrodes, standardized *IS-1* connections, are embedded in epoxy or polyurethane [101]. Figure 4.23 shows the interior layout of a DDD pacemaker

Tissue	42 MHz		64 MHz		900 MHz		1800 MHz	
	σ	ε_r	σ	ε_r	σ	ε_r	σ	ε_r
Air	0.000	1.000	0.000	1.000	0.000	1.000	0.000	1.000
Aorta	0.400	78.530	0.429	68.640	0.696	44.770	1.066	43.340
Bladder	0.282	27.290	0.287	24.590	0.383	18.940	0.535	18.340
Blood	1.183	101.200	1.207	86.440	1.538	61.360	2.043	59.370
Blood Vessel	0.400	78.530	0.429	68.640	0.696	44.770	1.066	43.340
Bone (Cancellous)	0.151	35.280	0.161	30.870	0.340	20.790	0.588	19.340
Bone (Cortical)	0.056	18.660	0.060	16.680	0.143	12.450	0.275	11.780
Bone Marrow	0.019	8.394	0.021	7.210	0.040	5.504	0.068	5.372
Brain (Grey Matter)	0.464	123.000	0.511	97.430	0.942	52.730	1.391	50.080
Brain (White Matter)	0.260	83.570	0.292	67.840	0.591	38.890	0.915	37.010
Cartilage	0.433	73.300	0.452	62.910	0.782	42.650	1.287	40.220
Cerebellum	0.648	156.200	0.719	116.300	1.263	49.440	1.709	46.110
Cerebrospinal Fluid	2.034	103.000	2.066	97.310	2.413	68.640	2.924	67.200
Colon	0.602	112.600	0.638	94.660	1.080	57.940	1.576	55.150
Fat	0.034	7.215	0.035	6.506	0.051	5.462	0.078	5.349
Gall Bladder	0.934	93.050	0.966	87.400	1.257	59.140	1.642	58.210
Gland	0.765	81.480	0.778	73.950	1.038	59.680	1.501	58.140
Heart	0.632	127.500	0.678	106.500	1.230	59.890	1.771	56.320
Kidney	0.682	146.300	0.741	118.600	1.392	58.670	1.950	54.430
Liver	0.414	96.220	0.448	80.560	0.855	46.830	1.289	44.210
Lung (Deflated)	0.507	86.450	0.531	75.280	0.858	51.420	1.279	49.380
Lung (Inflated)	0.274	44.970	0.289	37.100	0.457	22.000	0.637	20.950
Lymph	0.765	81.480	0.778	73.950	1.038	59.680	1.501	58.140
Muscle	0.671	81.260	0.688	72.230	0.943	55.030	1.341	53.550
Nail	0.056	18.660	0.060	16.680	0.143	12.450	0.275	11.780
Nerve	0.290	65.740	0.312	55.060	0.574	32.530	0.843	30.870
Oesophagus	0.858	98.010	0.878	85.820	1.187	65.060	1.698	63.230
Skin (Dry)	0.384	120.300	0.436	92.170	0.867	41.410	1.185	38.870
Skin (Wet)	0.457	91.610	0.488	76.720	0.845	46.080	1.232	43.850
Small Intestine	1.534	150.200	1.591	118.400	2.165	59.490	2.696	55.900
Spinal Chord	0.290	65.740	0.312	55.060	0.574	32.530	0.843	30.870
Spleen	0.692	139.900	0.744	110.600	1.273	57.180	1.780	53.850
Stomach	0.858	98.010	0.878	85.820	1.187	65.060	1.698	63.230
Tooth	0.056	18.660	0.060	16.680	0.143	12.450	0.275	11.780
Trachea	0.512	67.260	0.528	58.890	0.771	42.010	1.114	40.510
Uterus	0.881	111.000	0.911	92.120	1.270	61.120	1.764	58.940

Table 3.4.: Dielectric properties (conductivity σ and permittivity ε_r) for selected tissue types at 42, 64, 900 and 1800 MHz, based on Gabriel et al. [95].

Material	Permittivity ε_r	Conductivity σ [S/m]	Density ρ [kg/m^3]	Heat Capacity c [J/kg$\cdot$K]	Thermal Conduct. k [J/W$\cdot$K]
Air	1	0			
PEC	–	∞			
MP35N	–	968054	8430	754	11.2
Silicone Rubber	3	1e−13	1200	1000	0.22
Temperature probes	3.2	0	1000	800	1
Plexiglas	2.5e−3	2.6	1400	1000	0.2
Saline (0.9%)	77.8, 78	0.6, 0.78, 0.85	993	4200	0.62
Gel	56	0.82	1050	3200	0.42

Table 3.5.: Dielectric and thermal properties of used non-organic materials [96, 97, 32, 98].

Medium	Refractive Index $n = \sqrt{\varepsilon_r \mu_r}$	42 MHz [m]	64 MHz [m]	128 MHz [m]
Vacuum	1	7.1429	4.6875	2.3438
Saline (0.9% NaCl)	8.8318	0.8088	0.5308	0.2654

Table 3.6.: Wavelengths in media [99, 100].

system (*Advisa* by Medtronic).

To differentiate various pacemaker types, the *NBG* (NASPE/BPEG Generic Pacemaker Code; NASPE: North American Society of Pacing and Electrophysiology, BPEG: British Pacing and Electrophysiology Group) naming scheme was created. The five digits are used in the following way

1. Place of stimulation (A: atrium, V: ventricle, D: atrium and ventricle, S: one place but not yet determined)

2. Place of sensing (0: none, A: atrium, V: ventricle, D: atrium and ventricle, S: one place but not yet determined)

3. Stimulation mode (0: none of the following types, T: triggered by atrium activity, I: inhibited/suppressed by heart activity, D: triggered and inhibited)

4. Adaption to stress (0: not available, R: rate adaptiv) [optional]

5. Multifocal stimulation (0: none, A: multifocal stimulation in the atrium, V: multifocal stimulation in the ventricle, D: A and V) [optional]

To adjust the control unit to the special requirements of a certain patient, it can be programed transcutaneously by a device placed on the chest during and after the implantation. For cases of malfunction, pacemakers have to provide a feature called *reed switch*. When the patient or another person places a permanent magnet above the control unit, the pacemaker switches into a V00 or A00 mode depending on the implanted electrode configuration. This magnet driven switch can also be activated accidentally by the static B_0-field of an MRI device (see section 2.1).

3.8. Pacemaker Leads

Compared to pacemakers, the variety in mechanical design between different leads is considerably larger. In cooperation with *Dr. Osypka GmbH*, an overview of commonly used mono- and bipolar pacemaker lead types was developed. They also provided the custom-made systems described in section 4.2.8.

Electrically, there are two main types of pacemaker leads: monopolar and bipolar (see

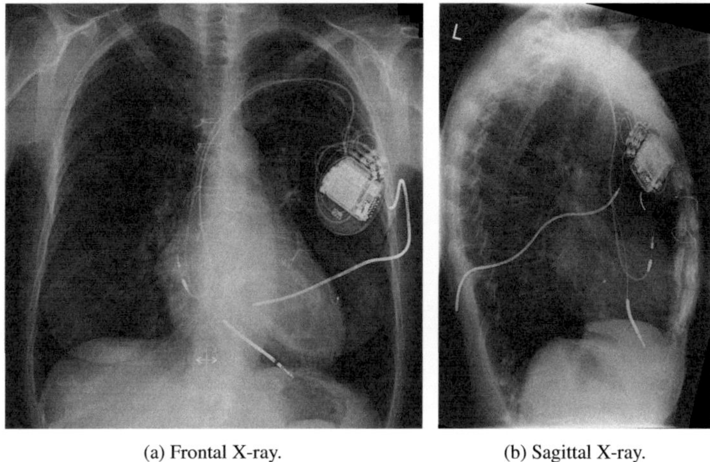

<table><tr><td>(a) Frontal X-ray.</td><td>(b) Sagittal X-ray.</td></tr></table>

Figure 3.11.: X-ray images of a patient with implanted cardiac pacemaker.

figure 3.12). In a monopolar system, the housing is part of the stimulation circuit as it acts as a counter electrode to the one at the tip of the lead. In a bipolar system, the stimulating field is only created between the two electrodes at the tip. This type of pacing is used in the vast majority of implanted systems today because it provides a much better control of the current distribution. Additionally, due to the smaller extension of the stimulating field, the bipolar method is more energy efficient. Because it is not required for stimulation, in bipolar systems the housing is said to have no galvanic contact to the circuitry and, as a consequence, neither to the electrodes at the tip anymore.

The pacemakers are implanted either in the left or right pectoral region (see figure 3.11). The attached leads preferably enter at the cephalic vein, otherwise later at the subclavian vein and follow the superior caval vein to the right atrium and if necessary further to the right ventricle. Nordbeck et al. provided a study of X-ray images illustrating occurring leads placements (see figure 3.13) [102]. Due to anatomical conditions, the position of the pacemaker and, more important, the course of leads is varying widely. This also determines if a chosen lead has to be coiled up close to the pacemaker housing or will be lain out flat without curls.

After reaching the intended position, the lead has to be fixed to the endocardium. Several types of fixation exist that are designed for specific surfaces. The helix type as

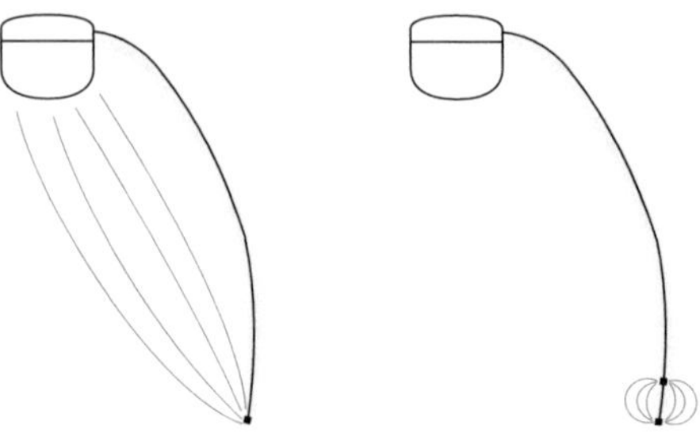

Figure 3.12.: Current paths for mono- (left) and bipolar (right) pacemaker systems.

shown in figure 3.14b, 3.14c and 4.24 provides a durable fixation in the smooth wall of the atrium. In the ventricle, the wall is covered with trabecula – they offer a stable support for leads with tines (see figure 3.14a). While the tines consist of non-metalic materials like silicone rubber, the helix is made of metal. The latter is connected with the wires and forms a galvanic contact for induced currents with the surrounding tissue. This is of special interest as the helix is buried in the myocardium and therefore excluded from constant heat dissipating blood flow. As a consequence, this type of electrode is of great importance when selecting leads for simulations.

The lead type shown in figure 3.14d is not placed inside the heart but will be attached outside on the epicardium. One application is bi-ventricular pacing.

In early lead models, the wires connecting the electrodes with the electric circuit were simple straight wires. Because of the constant mechanical stress caused by the heart's motion, many leads broke prematurely. In today's models the wires are coiled. This distributes the bending forces on many small locations and reduces the risk of broken leads significantly.

The link between lead and pacemaker device is established with a standardized *IS-1* connector. It allows an interoperability of leads and pacemakers of different producers. While the isolation of the leads consists of silicone rubber or polyurethane, the wires are made of *MP35N* or *Elgiloy*. The materials commonly used for the electrodes are platinum or iridium oxide.

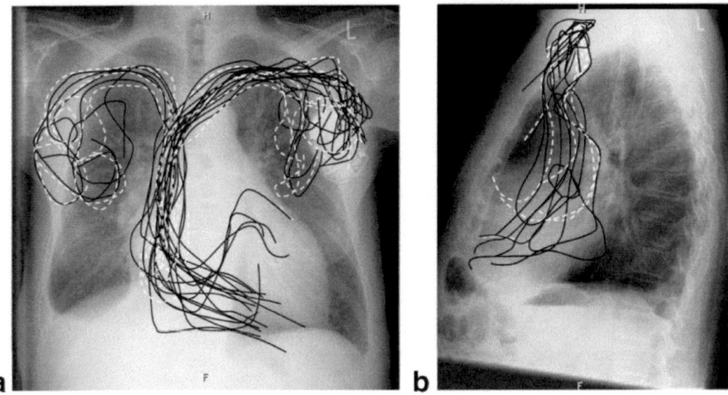

Figure 3.13.: Possible pacemaker and lead implantation sites identified by an X-ray image survey, (a) frontal view, (b) side view [102].

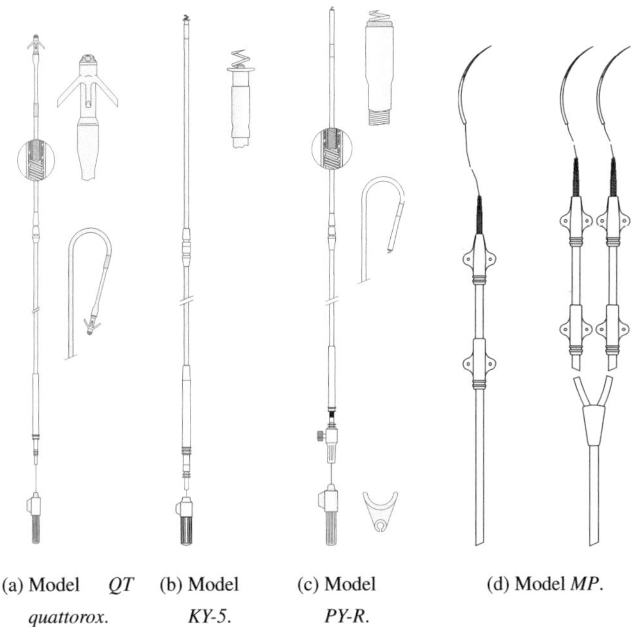

(a) Model *QT quattorox*.

(b) Model *KY-5*.

(c) Model *PY-R*.

(d) Model *MP*.

Figure 3.14.: Various types of pacemaker leads (Dr. Osypka GmbH).

4. Methods

In this chapter the methods used to carry out computer simulations and in-vitro experiments are presented. The first part will illustrate the software as well as the developed and used computer models, the approaches and assumptions, where necessary. The second part will describe the setups and systems utilized for the in-vitro experiments.

4.1. Computer Simulations

4.1.1. Simulation Environment

All numerical simulations described in this work were carried out using the commercial software package *SEMCAD* by Speag. It employed the FDTD technique described in section 3.2.4 to numerically determine electro(quasi)static, magneto(quasi)static as well as electromagnetic field distributions. Furthermore, it contained a thermo-solver based on the Pennes Bioheat Equation that allowed the determination of temperature changes based on previous EM simulations.

SEMCAD provided a basic CAD interface (see figure 4.1) that was used to create computer models of the phantom, leads etc. based on geometric primitives, e.g. cylinders, cubes and spheres. Combined with operations like subtraction, addition and extrusion all necessary geometries could be generated.

Beside the intrinsic model creation, the software also supported the import and export of external designs through filters for ACIS, SAT and AutoCAD. Another very useful feature was the scripting interface. Based on the language *Python*, SEMCAD provided an API (Application Programming Interface) to access nearly all elements and parameters of model creation, simulation settings, result extraction and post-processing of the results. Though incompletely documented, it allowed a fast way to automatize recurring tasks.

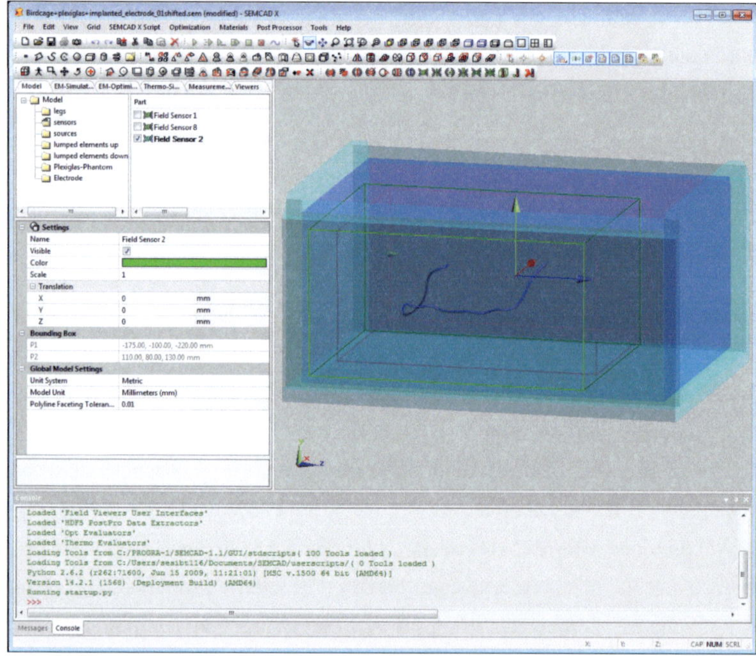

Figure 4.1.: CAD interface for modeling 3D objects in SEMCAD.

4.1.2. Preparation of Simulation and Simulation Environment

The preparation of the computer simulation required the consideration of several aspects. When occurring, metal objects were specified as *Perfect Electric Conductor* (PEC). The assumption underlying this type of material modeling is that for higher frequencies the EM fields inside a metal are predominantly located at the surfaces due to the skin effect (see section 3.2.1).

Based on this, in an object parametrized as PEC, the EM field calculation is concentrated to the surface layers. To evaluate the effect of this simplification, a setup was calculated with a metal object first set to PEC and then to metal object with corresponding conductivity values of *MP35N*.

After the geometric composition of a numeric simulation has been setup and all objects have been assigned with their dielectric properties, the next step is to define a grid for the later voxeling process. The grid has to provide an appropriate local resolution to represent all necessary details and this brings up the problem of computational cost. The complexity of the set of linear equations used in FDTD is anti-proportional to the minimum grid resolution. The finer the grid, the larger the equations that have to be solved. Dividing the mesh step size by two increases the computational cost in an equidistant grid by the factor of 16 [68].

When a small object in the center of the calculation domain should be discretized with a very fine mesh, in FDTD the same resolution will be used up to the boundaries of the calculation domain (see figure 4.2). The fine structures of the pacemaker leads collide with the large outer dimension of the field generating RF coils. The first requires a mesh step of 0.1 mm, while the second could be meshed with a one to several millimeter grid resolution.

Two approaches were used in this work to overcome this dilemma: Hardware Acceleration (see section 3.2.5) and the *Huygens box* feature of SEMCAD described in section 3.2.4.

The workstation used for the presented work could handle simulation sizes up to 24 GB and process about 40 MCells per second. For many of the simulations this would have meant several weeks of computation for one single setup though. Due to a parallel project done at the IBT, a workstation with an attached hardware acceleration unit could be used. The *CIB 1000* system by Speag contained two NVidia *Quadro FX 5600* boards

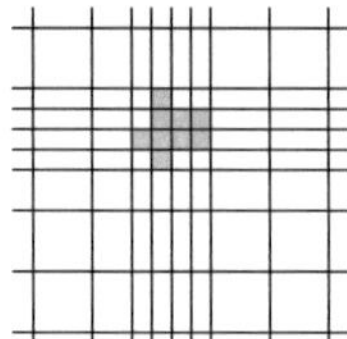

Figure 4.2.: Effect of local high mesh resolution on global grid.

with 1.5 GB memory per GPU in an external chassis and two AMD *Dual-Core Opteron 2214* CPUs accompanied by 16 GB RAM in the main housing. The system offered a processing speed of up to 580 MCells per second, reducing the computation time for optimal cases by the factor of 14.5. The speed up for an actual simulation was determined by the complexity of the simulation (number of mesh cells) and the parameters of the included materials. A setup with a lot of metal like objects (parametrized to PEC) required more iterations until steady state criteria were met than a configuration with less. Due to the limited amount of on-board memory of the GPU units, simulations with too many cells caused the solver to distribute the workload between GPUs and CPUs leading to a significant drop in speed because of the different performances of GPU and CPU. The management of the employed resources including the distribution of the workload between GPU and CPU happened transparently for the user.

For applications in which the single precision solver could not achieve an appropriate steady state, a double precision option was available in SEMCAD. For the EM simulations it was never necessary, only some of the temperature simulations had to be calculated using this feature.

4.1.3. Anatomic Voxel Models

The basis for computing realistic electromagnetic field distributions in a virtual environment that are comparable to the ones inside the human body requires detailed anatomical models. They provide the link for every volume element inside the model to the dielectric properties of the respective tissue types. The generation of those models is a complex, time consuming and expensive process, hence only few of them exist.

The most commonly used ones are derived from the *Visible Human Project* [103]. Based on their available MRI and photographic datasets, a slice-based segmentation was done for the whole body of a male and a female volunteer. For every element of the resulting mesh, the type of tissue was determined and marked with an index for later reference to dielectric properties (see table 3.4 for values). The dataset had been down-sampled for resolutions of 2, 3, 4, 5, 6, 8, 10, 12 and 15 mm to allow simulations with reduced complexity.

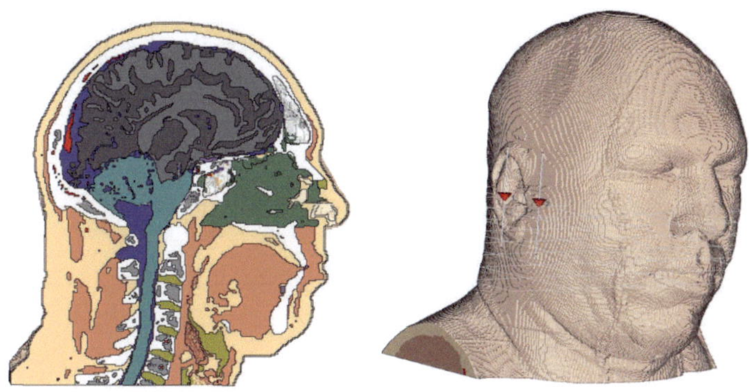

(a) Sagittal section showing segmented tissue types.

(b) Perspective view of head region.

Figure 4.3.: Fully segmented Visible Human dataset (1 mm resolution in x,y,z direction).

A different implementation was chosen by IT'IS Foundation for their *Virtual Family*. While also based on MRI images, the final dataset is vector based on every slice and therefore avoids the problem of stair-casing artifacts in this slice. Because of its representation as curves, the data can be better interpolated for higher resolution computations compared to a classic voxel dataset. The complete set is shown in figure 4.4a. It comprises

- a 34-year-old male adult

- a 26-year-old female adult

- a 14-year-old male child

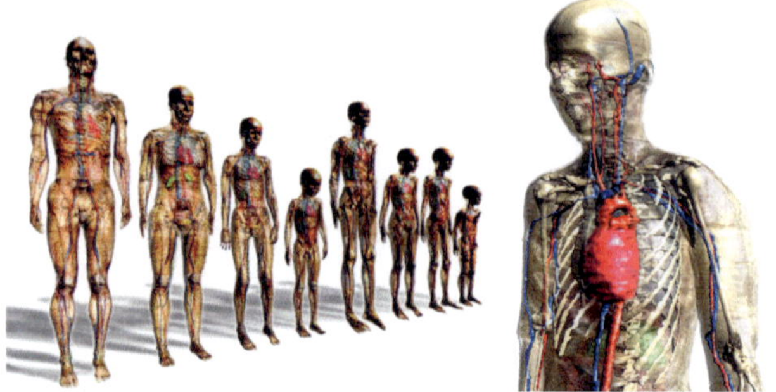

(a) *Virtual Family* and *Virtual Classroom* mod-(b) Dataset of 26 year old fe-
els.   male.

Figure 4.4.: Fully segmented MRI-based datasets by IT'IS Foundation.

- an 11-year-old female child

- an 8-year-old male child

- an 8-year-old female child

- a 6-year-old male child

- a 5-year-old female child

Each of them was segmented differentiating between 84 tissue types and organs.

4.1.4. 3D Surface Models

As described above, the generation of detailed volumetric anatomical datasets is expensive in cost and time. When doing numeric whole-body SAR evaluations, a full featured model including all possible details is not always necessary.

As a potential alternative, a new approach to create anatomical models was evaluated: three-dimensional whole-body surface scans. Initially, a local system was intended, but proved to be too expensive given the few number of expected scans. In cooperation with *Deutsche Sporthochschule* (Cologne, Germany) it was possible to use their laser-based system. The Vitronic *Vitus XXL* (see figure 4.5) provided a spatial resolution of 1 mm in x and y plane and 5 mm in slice distance in z direction.

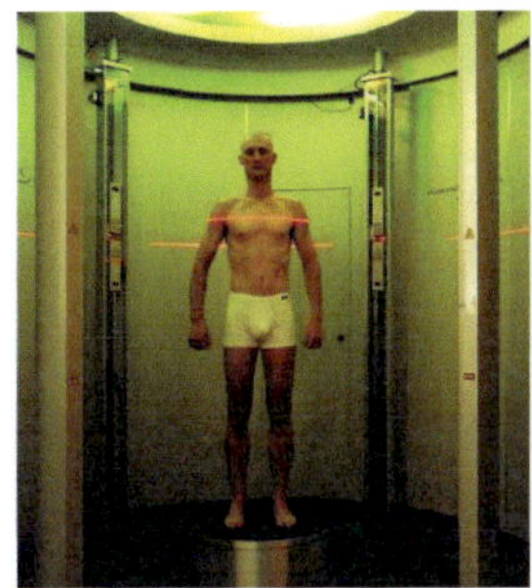

Figure 4.5.: 3D Laser Body Scanner (Vitronic).

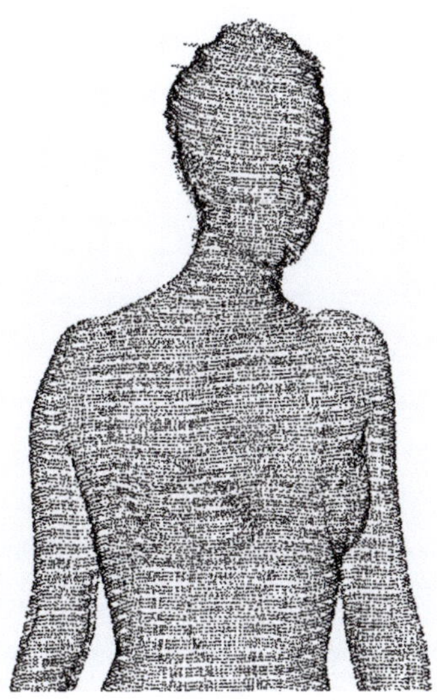

Figure 4.6.: 3D scan, raw point cloud data, 90.038 points.

The recording time for the scan was around 13 seconds and therefore significantly faster than a MRI procedure. The scanner's output consisted of a point cloud dataset. The raw data contained several types of artifacts: areas of the torso that had been shielded by the limbs could not always be completely detected by the laser. The same happened on parts where the laser could not be properly reflected. One reason for the last described problem were surfaces parallel to the laser-light like the top of the head or the shoulders. Another were rough surfaces like hair or ear where the light was more scattered than reflected. Those areas required subsequent post-processing.

It was performed with the software *Studio 10* and *Studio 11* by Geomagic. The software allowed the reconstruction of the missing surfaces, de-noising of rough areas and finally the transformation into a closed surface 3D STL dataset. This was important, because EM simulation tools require an intact surface to be able to attribute the volume with dielectric properties.

4.1.5. Generic Organ Models

Since the afore described surface scans did not provide any information about the circumstances inside the shell, a method was required to attribute the volume with properties resulting in realistic field distributions. The initial approach was to fill the shell with one material, whose dielectric properties result in an EM field and/or SAR distribution that is comparable to a full-featured model. But several publications have shown the importance of a model design that at least contains representations for the most important organs [104, 105, 38]. The reasons for this are discontinuities in the conductivity and permittivity that can lead to reflections and standing waves. As a consequence, local superelevations occur that would not appear in a homogeneous model.

Since the growth of organs is neither constant nor isotropic and varies for different types of organs (see figure 4.7), a method was developed to generate generic organ models matching in age and gender to the current surface scan. Derived from one exemplary fully segmented dataset, the organs were deformed according to the specifications of the proband. The basis for the deformation algorithm were patterns published by Cristy in 1980 who described how organs change in volume and shape during growth [106, 107]. In combination with reference organ weights published by the International Commission

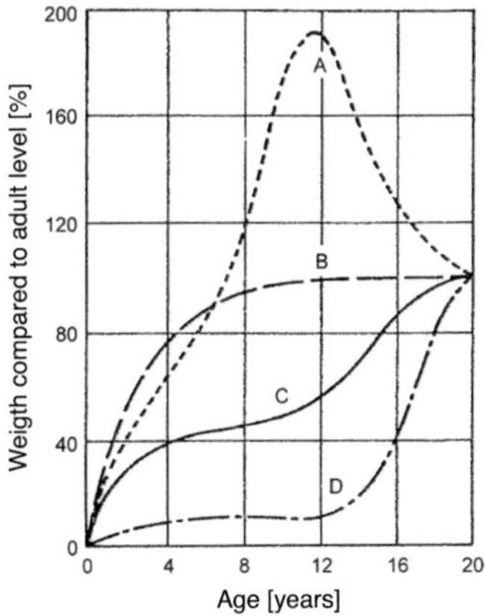

Figure 4.7.: Four different groups in organ growth: A = lymphoid, B = neural, C = general, D = reproductive organs.

on Radiology Protection (ICRP), the organ models were scaled to match any arbitrary gender and age.

After segmentation, the volumetric organ models had to be aligned to the mathematical models, because the later transformation incorporated 3 orthogonal scaling parameters, each specific for a spatial direction. This alignment was achieved using principle components analyses and is illustrated with the liver dataset in figure 4.8. Subsequently the transformation was applied and the resulting organ models were verified by comparing them with examples from the SPEAG *Virtual Family*.

4.1.6. CAD Models

To analyze a specific configuration of a phantom together with an object of interest inside a body coil, the arrangement had to be modeled as a three-dimensional representation in the computer.

Straight Wires

A prerequisite for the examination of implants exposed to electromagnetic fields is an appropriate adaption of the implant to achieve reliable results. The first idea could be to simply take highly detailed pacemaker and lead models and place them in a voxel

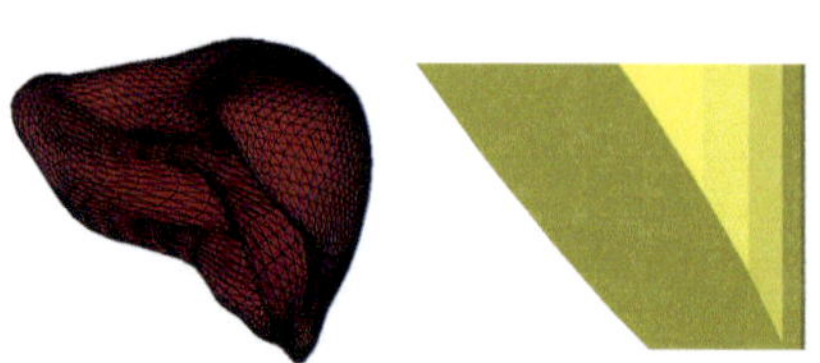

(a) Comparison of real and mathematical liver phantom.

(b) Aligned real and mathematical liver phantom.

Figure 4.8.: Mathematical and segmented model of real human liver.

dataset. Unfortunately because of the complexity of these models, the simulations quickly become impossible to be solved due to limited computing resources. Furthermore, the role of patterns in heating like position or orientation could be difficult to differentiate from effects caused by the circuitry inside the devices. Therefore, a model commonly used in the literature [97, 108, 30] for simulations and in-vitro experiments are (isolated) wires.

Because of the plain structure, it is well suited for examining heating patterns in respect to position and orientation. Since the wires do not contain any passive or active elements, all measured or simulated temperature changes can solely be attributed to influences of external factors like the surrounding fields, position etc.

A matrix of different lengths and positions was created, each with a diameter of 1.5 mm and a coating of 1 mm. The dielectric properties of the isolation were defined according to silicone rubber (see table 3.5), the leads were modeled as PEC. Although the diameter of the core and especially the isolation are bigger than regular leads, this was chosen as a trade off between accuracy and simulation complexity. The length of the wires was set to 160 mm. The position of the wires was chosen as the parameter for a simulation study and varied between 30, 50, 70, 90, 110 and 130 mm off the iso center of the body coils. The impact of defining the metal parts to PEC and not to real-world values like for example Elgiloy or MP35N was examined separately.

Pacemaker Leads

As illustrated in section 3.8, of the numerous pacemaker lead types available types only a few are commonly implanted in patients in daily clinical practice. Three types were selected and transformed into computer models for later simulations:

- Bipolar *CapsureFix Novus* with helix fixation (Medtronic)

- Bipolar model with helix fixation (Osypka)

- Bipolar model with hook fixation (Osypka)

The miniaturized structures at the tip (ca. 0.2 mm) in contrast to the long extension of the supply line (lengths of commonly implanted system: 45, 56, 65 cm) caused problems when meshing and generating the mesh. As a workaround, the following assumption was made: the supply line could be truncated without influencing the heat dissipation pattern at the tip qualitatively. Nevertheless, the length of the lead indeed can have an additional significant impact on the amplitude of occurring heating. This aspect had been already extensively analyzed in several studies for example by Mattei [30].

Pacemaker Lead Following Anatomical Structure

In the previously described sections, the leads and wires were always laid out flat in one plane. This approach was chosen, because it made identification of patterns in heating easier. But when a lead is normally implanted in a human being, it has to follow the anatomical structures that determine its path from the entry point in the venous system down to the heart (see figure 3.13). This path is normally much more winded and complex and not flat at all. Only the remaining part of the lead is curled up next to the pacemaker housing where it can have a more or less flat orientation.

To adapt those circumstances, a lead model was designed that followed the venous system of the anatomical voxel dataset (see figure 4.9). It entered the vessel at the clavicle, continued in the superior caval vein to the right atrium and ended at the bottom of the right ventricle. The metal core had a diameter of 1.4 mm and was coated with a 1 mm thick layer made of silicone rubber. On both ends, the isolation is cropped for 2 mm. Initially the voxel dataset should not only serve for developing the wire model but also to provide the surrounding for the simulation. But due to issues at the boundaries with

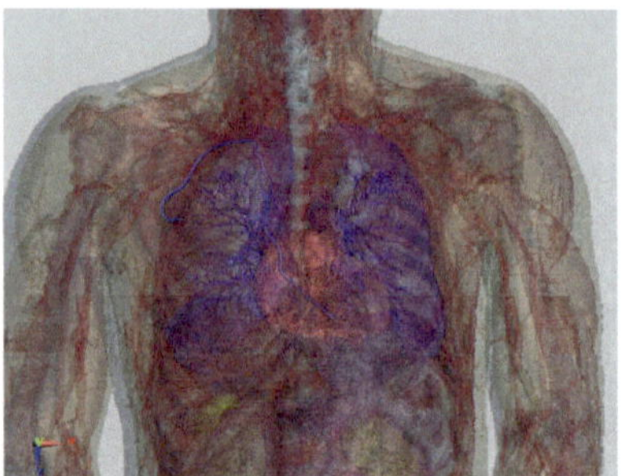

(a) Voxel dataset with all tissue classes (frontal).

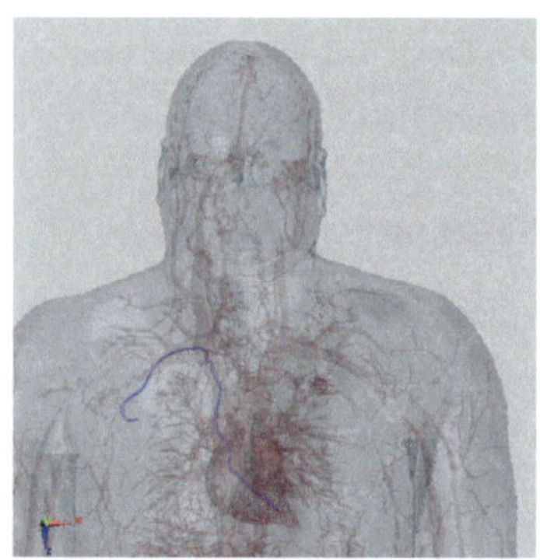

(b) Voxel dataset with skin and vessels (frontal).

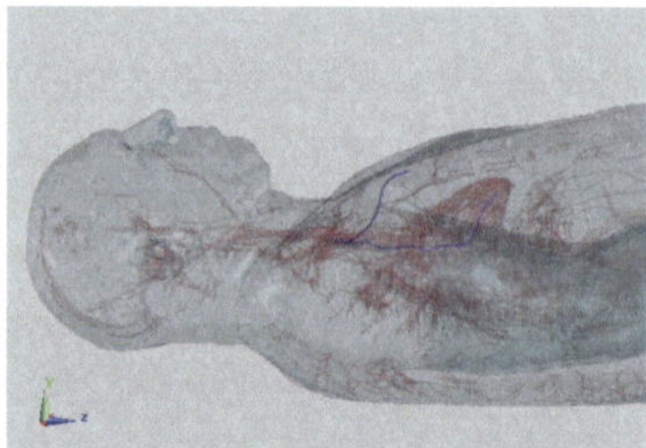

(c) Voxel dataset with skin and vessels (side view).

Figure 4.9.: Computer model of lead following the venous system inside the anatomical voxel dataset.

the voxel dataset, the therefore necessary extension of the Huygens box collided with the minimum mesh step required by the wire model. Hence, the wire was placed in the regular Plexiglas phantom to determine, whether the anatomically correct lead path would produce significantly different SAR distributions than a straight wire.

Pacemakers

The wire models described above meant a greatly simplified approach compared to the complex structure of a normal pacing lead. Because of the unknown interior of the pacemakers, the simulation setup was based on the measurements of an opened device. The acquired values for the input impedance were implemented in the simulation as lumped elements connecting the proximal end of the lead with the housing. For the simulation in the birdcage, a series connection consisting of an inductance and a resistor was parametrized with extrapolated values at 64 MHz. To determine the impact of the type and dimension of the used lumped elements, a second setup with a single capacitor configured to 5 nF was used (see figure 4.10c).

As part of this project, possible workaround to reduce induced currents by modifications to the pacemaker/lead system were to be evaluated. One for example could be the inclusion of a conductor in the lead to attenuate RF currents. To test the efficacy of this approach, the conductor in the setup described above was additionally set to an exceptional high value of 172.6 μH (see figure 4.10b). In all cases, a wire coated with silicone rubber (length 16 cm, 2 mm bare ends) served as lead model. The pacemaker housing consisted of a 1 mm thick PEC shell with chamfered edges as in regular devices. In the numerical simulations, this helped to avoid artificial peak distortions in the EM field distribution. The outer dimensions were adapted from a Medtronic *Adapta* dual chamber system and a St. Jude *Frontier II* device and set to $5.5 \times 5 \times 7$ cm. The inner volume was left empty and attributed with the dielectric properties of vacuum.

An early version of the pacemaker model included a silicone shell for the housing as well but was later discarded. It seemed that even dual bipolar systems require a metallic contact between housing and surrounding tissue, although it is not needed it for the actual stimulation.

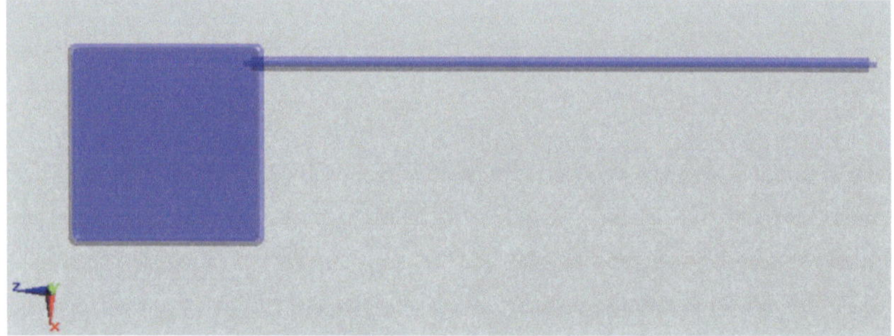

(a) Pacemaker with lead.

(b) Detail view of lead connected to housing via two lumped elements: Resistor (upper) and inductor (lower).

(c) Detail view of lead connected to housing via two lumped elements: Capacity.

Figure 4.10.: Computer models of pacemaker and lead with lumped elements.

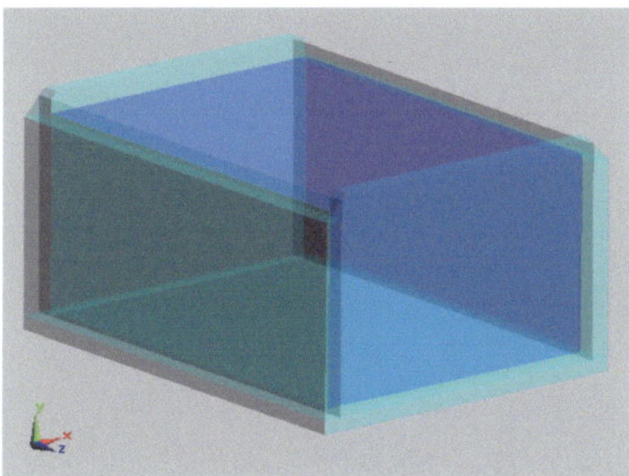

Figure 4.11.: Model of plexiglas phantom.

Plexiglas Phantoms

All simulations were carried out with the same type of Plexiglas phantom. It was an exact replication of the first in-vitro phantom (see figure 4.11 and section 4.2.3), but with the difference that the inside walls had been removed to allow surveys with objects freely shifted around in certain areas. They were not required for stability in a computer model and due to the massive structures of the walls, they collided with the tested objects.
The fixtures for the leads and the fiberoptic temperature probes were omitted as well. Though unavoidable in the in-vitro experiments, they implicated two problems. They also consisted of Plexiglas which is an isolator and therefor disturbs current and EM field distributions. Secondly, the temperature dissipation is altered because of the lack of convection since Plexiglas is a solid. The walls were parametrized with the dielectric properties of Plexiglas, the free volume inside was configured as saline.

Birdcage Body Coil

The model of the birdcage coil was provided by Speag, but for completeness, the principle functionality should be described here. The coil consisted of two times 16 T-shaped elements called legs that were arranged to form a cylinder (see figure 4.12). As material PEC had been chosen. At the junction of always two lower ends, an edge source was placed and configured to deliver a current of 1 A with a frequency of 64 MHz corresponding to the Lamor frequency at 1.5 T. The stimulating signal was delayed for the

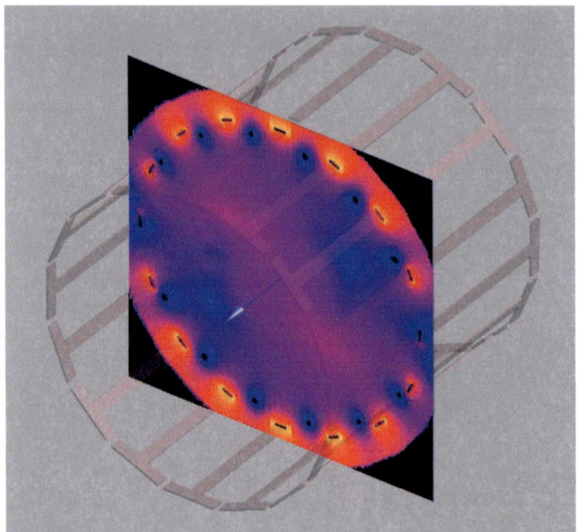

Figure 4.12.: Model of birdcage body coil with cross section of B_1-field.

second leg pair by 1/16 of a period, for the third by 2/16 etc. to initiate a rotating field. On each side of the coil the top ends were coupled pairwise by an R-C combination with values of $10\,\Omega$ and $63\,\mathrm{nF}$ respectively. Those elements helped to tune the coil and to maintain the resonance of the system.

The diameter of the coil was 63 cm and the length 65 cm. The volume inside providing a homogeneous field distribution with a magnetic field intensity of around $2\,\mu\mathrm{T}$ was large enough to comfortably place both types of Plexiglas phantoms.

Open MRI Body Coil

The design of the open MRI coil was initiated by a presentation of Kleihorst (Philips Medical Systems) and Vogel (Ansoft Corporation). Since details regarding the implementation were undisclosed, the only adapted aspect was the shape of the two discs placed above and below the patient table (see figure 4.13a). The silhouette of open MRIs was also indicating such a shape.

The dimensioning of the two discs was as well inspired by the outer dimensions of a real device and the specifications given in an installation planning guide for a Philips Panorama. The upper and the lower magnet covers had a distance of approx. 44 cm, the horizontal free space around the field of view was about 1.5 m. Based on these values, the discs were placed with a distance of 45 cm and had a diameter of 1 m and a height of 5 mm. The tines were 6.5 cm long, 3 cm wide and set to the same height as the discs: 5 mm. Each item featured a small gap where later the edge source could be placed. Furthermore, all tines were rotated by $45°$ towards the outward facing side of the discs. Because of the large number of objects to create (final version: 128 tine elements, 64 lumped elements) – all with different parameters – the whole coil was created with a Python script. It alleviated the development of the coil significantly, because it enabled the testing of different configurations in a reasonable amount of time. The number of necessary tines was identified by testing configurations with 4, 8 and 32 copies. Only with the last number of tines, the field was sufficiently homogeneous.

The difficult task then was to develop a mechanism of exciting the discs in a way that would result in a rotating B-field between the discs. Furthermore, a very good homogeneity throughout the field of view (FOV) had to be ensured. Finally, the B-field intensity should be comparable to the field inside the birdcage coil.

The basic principle of exciting the coils was comparable to the the birdcage coil, since here a rotating field should be created as well. As shown in figure 4.13c, an edge source (type: current, amplitude: 1 A, frequency: 42 MHz) was placed in the gap of each tine around the disc.

The stimulation protocol and method was determined in combination with the tine number. The first idea was to stimulate the tines pairwise, to establish a current path between the points where the current entered the disc and to subsequently create a corresponding magnetic field between the two discs. This only worked properly for configuration with up to four tines. When the number was increased, the fields got inhomogeneous and the rotation collapsed prematurely. The second attempt was to stimulate with a phase delay proportional to the number of connectors like in the birdcage coil. Together with a contrary polarity of the signal in the opposite disc, this finally generated a rotating, adequately homogeneous field with an intensity comparable to the birdcage coil.

The quality of the generated field will be demonstrated with a comparison of B-fields present in both coil types.

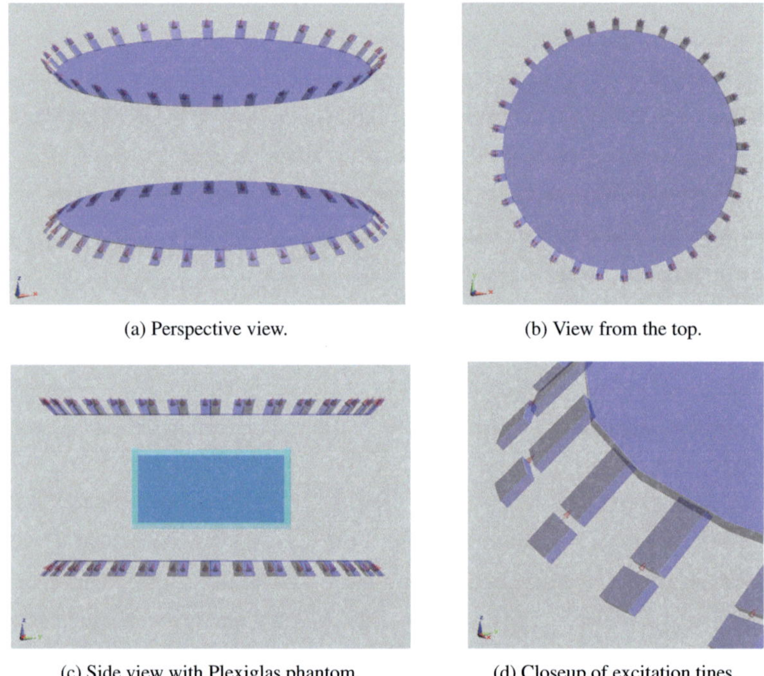

(a) Perspective view.

(b) View from the top.

(c) Side view with Plexiglas phantom.

(d) Closeup of excitation tines.

Figure 4.13.: Body coil model for open MRI application.

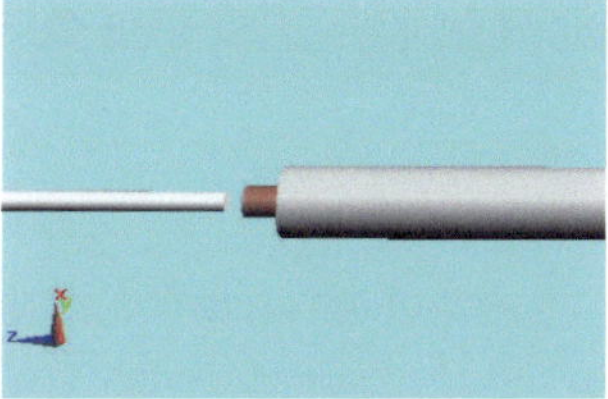

Figure 4.14.: Computer model of wire with adjacent fiber-optic temperature probe.

4.1.7. Influence of Temperature Probes on the Electromagnetic Field Distribution

When executing in-vitro experiments, the temperature as most crucial information is commonly acquired with fiber-optic temperature probes. Though equipped with an only 1 mm thick tip, the probe is a foreign object with dielectric and thermal properties that are different from the surrounding tissue simulating liquid. By means of computer simulations, the impact of the probe on the electromagnetic field and subsequently the temperature distribution should be evaluated. In one of the simulations with isolated wires, a model of the probe's tip was placed opposite to the tip of the wire (see figure 4.14). The parameters used for the probe can be found in table 3.5.

4.1.8. Influence of Phantom Filling Parameters on Temperature Distribution

For realistic simulations, the filling of the phantom had to be attributed with appropriate dielectric and thermal properties as also covered later in the section on in-vitro experiments (see 4.2.4) In the numerical simulations, the parametrization of the filling was much easier, hence different configurations could be tested.

The dielectric as well as the thermal parameters for a saline filling could be directly derived as shown in section 3.6). To assess the changes induced by adding a gelling agent, a simulation was defined, where the filling was adjusted to values presented by Neufeld [32] and listed in table 3.5. Because the implementation of the Bioheat equation in SEMCAD only considered the conductive term and the energy deposited by an EM field, the results again will represent a worst case scenario.

4.1.9. Realistic MRI Sequences versus Scaled Continuous Wave Excitation

The EM field and SAR distributions determined by computer simulations were primarily defined by the course and the power of the excitation signals used to feed the coils. In a real MRI device, these also called *sequences* follow complex patterns to manipulate the spins. The sequences were especially designed for specific medical questions (see section 3.1.2). When trying to reproduce them for computer simulations, the problem arouse that they were only defined in principle and the actual implementation in the MRI devices was unknown. All information request to the manufactures produced only vague responses.

This initiated a simulation study that should evaluate whether a realistic implementation of the sequence was necessary at all to achieve realistic results. Two sequences were adapted for temperature simulations: saturation recovery and spin-echo. Like in a real MRI device, the RF pulses were generated by switching a continuous wave signal on and off. The chronological patterns of sequences were used to control this.

The principle is illustrated on the basis of the saturation recovery (SR) sequence. The dominating parameter of the SR sequence is the repetition time T_R. During the MRI procedure, the RF signal is repeated every T_R. For the simulations, T_R was set to 10, 20 and 50 ms. The stimulation was carried out for 10 s followed by a cooling-down period of 5 s. According to Dössel, this was in accordance with real MRI system [56].

The scaling of the continuous wave signal was computed as follows. The ratio of RF pulses and breaks could be determined with

$$c = \frac{\sum t_{RF\,pulse}}{T_R} \tag{4.1}$$

Since the power of the pulsed signal and the continuous one should be equal, the latter one could now be determined as

$$P_{cont} = c \cdot P_{pulsed} \tag{4.2}$$

4.1.10. Encapsulation of Pacing Lead

To analyze, whether fibrotic tissue would have any impact on induced currents, a simulation profile was created based on the published conductivity values of Grill et al. [43],

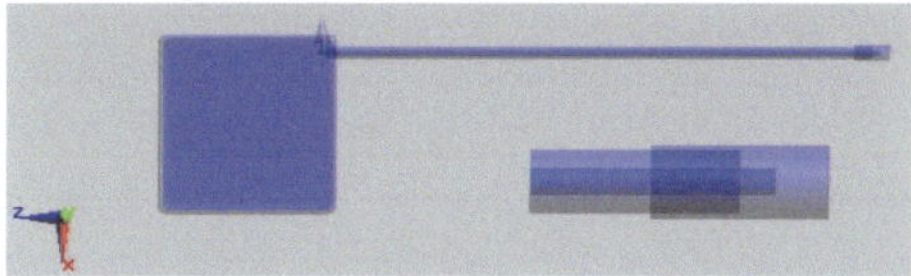

Figure 4.15.: Computer models of pacemaker and lead with encapsulated tip.

Angle	In birdcage	In open MRI
0°	regular	regular
45°	intermediate	intermediate
90°	like open MRI	like birdcage

Table 4.1.: Relation of different wire/lead positions in open MRI and birdcage coil.

who proposed 0.15 S/m for the fibrotic tissue. Since that publication did not cover eventual permittivity alteration, this parameter was derived as a hybrid value from heart muscle tissue (very well perfused) and cartilage (no perfusion). According to Gabriel, at 64 MHz heart tissue has a relative permittivity of 106 and cartilage one of 62.91 [93, 94, 95]. To emphasize the effect of a reduced perfusion, the simulation was parametrized with $\varepsilon_r = 70$.

As tested object the pacemaker/lead model described above was employed again. This time, the blank tip was additionally covered with a small cylinder attributed with the just derived dielectric properties (see figure 4.15).

4.1.11. Comparison of Effects of Open and Bore Hole MRI Coils

The hypothesis if an open MRI coil meant a lower risk for induced currents in the pacemaker leads was tested with the following simulation setups. A wire was firstly placed horizontally, then flipped upwards by 45° and finally fully upright in the phantom. Because the rotational planes of the B_1-field vectors in the birdcage and the open MRI coil showed an angle of 90°, a wire or lead in the 90°-position would be exposed to the field like in the opposite device type (see table 4.1).

In total, six simulations were done in the higher phantom with a 16 cm long metal wire coated with silicone and 2 mm bare ends on both sides. In all configurations, the wire was voxeled with a resolution of 0.2 mm. To evaluate the impact of the orientation,

SAR histograms for all configurations were extracted.

The same approach was later used in the in-vitro experiments, but in that case with a pacemaker and a lead.

4.1.12. Evaluation Methods

The evaluation of EM field simulations and temperature distributions around complex and fine structures can be difficult and potentially misleading. Putting the focus only on maximum E-field, current distributions and subsequently SAR or temperature values can wrongly overestimate aspects in the results. When compared to the surrounding tissue, current distributions in a metal object for example will naturally produce extremely high values on the surface mesh cells of the object because of the conductive nature of the material and the skin effect. So an evaluation that excluded the metal objects would represent the hazardous potentials considerably better.

Another difficult task is the assessment of effects on implant-tissue boundaries. For miniature structures like the helix shaped lead fixations, the common visualization methods like single layer maps or 3D iso surfaces make it hard to evaluate the whole volume and quantify portions affected by dangerously high SAR values.

In this work the results were therefore additionally presented as histograms for a certain volume around the lead tips. This also allowed the classification of few peak SAR or $\vec{J}$ values as insignificant artifacts or valuable information.

Commonly the SAR values derived from a computed E-field or current density distribution are averaged over 1 g or 10 g (see section 3.3). Because the corresponding volumes are rather large compared to the fine structures of the objects that are examined in this work, it should be evaluated, how the assessment is influenced by the averaging volume. Mattei had done similar research in 2009 and found a maximum deviation of up to 90 % between results averaged over 0.1 and 10 g [31].

4.2. In-vitro Experiments

The degree of maturity of numerical simulations has reached a level where the results in many cases can measure up with in-vitro measurements. Nevertheless, for a complete validation it is still common practice to test individual setups in in-vitro experiments. The following second part of this chapter will describe the setup of the experiments, the used MRI devices and sequences and the devices needed to acquire temperature maps and curves.

4.2.1. MRI Devices and Sequences

The experiments presented here were carried out in two different imaging facilities. The first was the cardiac MRI laboratory (Philips 1.5 T, bore hole type) at *Medizinische Klinik, Innere Medizin* at the University of Heidelberg, Germany. The second site was a private radiology practice also in Heidelberg equipped with a Philips Panorama 1 T (open MRI).

At both locations, a technician from Philips had installed special sequences that maximized the SAR. The reason is that the expected temperature elevations are small, so with a maximized SAR a signal above noise can be expected (see section 4.2.2). Due to the regulations for clinical devices, this was limited to 4 W/kg. The fact that the sequences really caused a substantial amount of RF energy could easily be asserted due to intermittent automatic cooling-down breaks. All in-vitro experiments were stimulated with the following protocol:

- Survey scan

- Reference scan (if necessary)

- An arbitrary number of T2 TSE sequences (parameters: SAR 4 W/kg; flip angle 180°; 2 durations: 3.5 min. and up to 28 min.)

- An arbitrary number of 3D 6TFE ct sequences (parameters: SAR 4 W/kg; flip angle: 70°; 2 durations: 4.5 min. and up to 26 min.)

The length of the T2 and 3D sequences could either be adjusted by using the long configuration or a combination of multiple shorter ones. When required, the heart rate triggering was bypassed with a simulated ECG adjusted to 60 bpm. Although not relevant for heating, in all sessions in the bore-hole device, a 32-channel coil received the signals.

4.2.2. Fiber-optic Temperature Probes

The measurement of temperature distributions and changes in MRI environments is commonly done with fiber-optic based systems. Since the RF fields during an imaging procedure also influence a thermometer containing metallic parts, only fiber-optic probes can acquire the temperature exactly and undisturbed.

The device used in this study was a *FOTEMP4* by OPTOcon (see figure 4.16). It provided four parallel measurement channels with an intra-measurement resolution of $0.1\,^\circ$C and a sample rate of 0.5 Hz. The control unit was located outside the MRI cabin, connected to 4 *TS2* probes with a tip diameter of 1.2 mm and a feed line length of 8 m. To protect the fragile cords, they had been inserted into a flexible tube that avoided sharp bending and, as a side effect, made handling and storing of the cords much more convenient.

The placement of the particular probes followed the same pattern in all experiments. They were positioned to read temperature at

- the electrode at the tip

- the ring electrode

- the pacemaker housing

- one more distant spot close to the wall of the phantom for reference.

The last position was chosen to detect a general temperature increase inside the phantom. This was part of every regular MRI procedure also without any implant present. The source of this heating were not the EM fields itself but the waste heat of the MRI device while generating the RF signals.

When placing the probes at the two electrodes, great care had to be taken in order to position the probes as close as possible. This was achieved with the probe/lead fixation ring shown in figure 4.21 and 4.20. As the results of a thermo-simulation showed, a small distance could cause a significant drop in the captured temperature values (see section 5.1.8).

The data was either available as a current signal for each of the channels (0–$10\,V_{DC}/4$–20 mA), or from a serial RS232 output connected via a RS232-USB converter (*FT232* by FTDI). The serial port allowed logging of the data on a computer with a small application

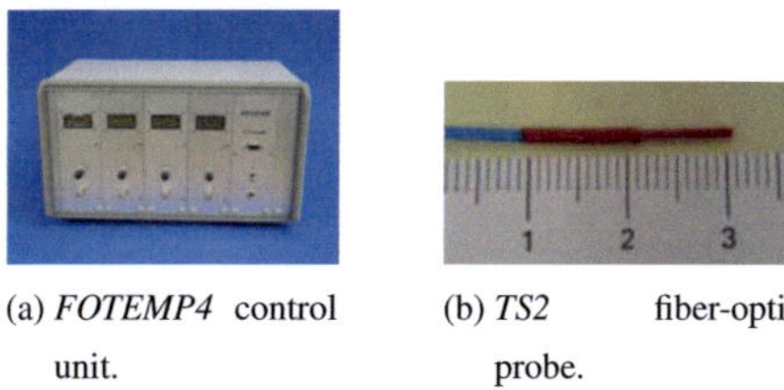

(a) *FOTEMP4* control unit.

(b) *TS2* fiber-optic probe.

Figure 4.16.: Temperature measurement system by OPTOcon (Images: OPTOcon).

```
18.04.2010 23:51:16,515   24,60   24,50   24,50   24,50
18.04.2010 23:51:16,781   24,60   24,50   24,50   24,50
18.04.2010 23:51:17,046   24,60   24,50   24,50   24,50
18.04.2010 23:51:17,312   24,60   24,50   24,50   24,50
18.04.2010 23:51:17,578   24,60   24,50   24,50   24,50
18.04.2010 23:51:17,843   24,60   24,50   24,50   24,50
```

Figure 4.17.: Data log of a sample MRI measurement session.

provided by the device manufacturer. It produced an output with a precise, absolute timestamp and temperature values for each of the 4 probes as shown in figure 4.17. The data was later analyzed using *MATLAB* (Mathworks).

The probes had already been calibrated by the manufacturer and labeled with an absolute accuracy of $\pm 2\,^{\circ}$C. To achieve precise results with the same reference value despite a possible total deviation of $4\,^{\circ}$C, the probes could be calibrated in the control software. Previous to each measurement session, all four probes were located at one small spot inside the phantom and adjusted to the average of all 4 temperature values. Unfortunately, on one channel this user calibration degraded over time. The failure was later repaired by the device manufacturer.

4.2.3. Phantoms

During the course of the in-vitro experiments, three different phantoms were used. All consisted solely of Plexiglas panels and plastic screws. The first two were derived from the phantom as described in ASTM standard F2182-02a (see figure 4.18 (a) for dimensions) [109]. This standard contains recommendations for measurements and testing procedure of medical implants in MRI environments. The outer dimensions follow the

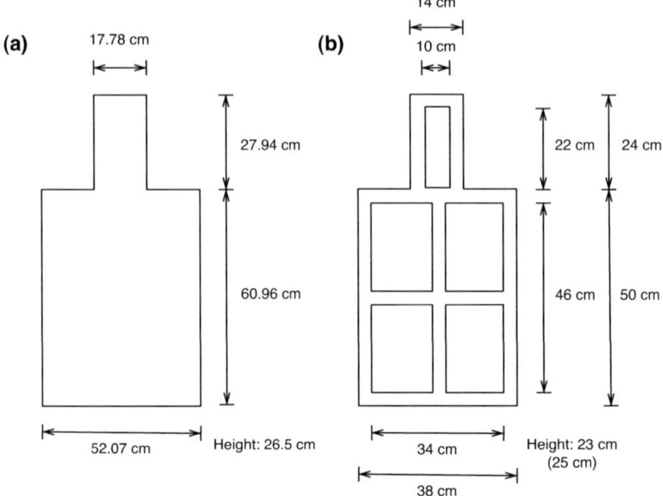

Figure 4.18.: Measurement phantom I: (a) as described in ASTM F2182-02a [109], (b) already available at IBT.

abstract form of a human torso and allow the placement of implants in positions similar to regular implantation sites.

For the comparative trials between open and bore hole MRI, a different basin for the saline was necessary because the phantoms were not high enough. It consisted of a plastic container with 2 mm walls and could be filled with 25 l of saline.

The second phantom was very similar to the first simply in addition the "head" part had been removed (see figure 4.19). The reason for this was that no testing needed to be done in the head region and because of several wholes in the "face" area prevented a complete filling with saline up to the top.

The design of the third phantom varied significantly from the first two: it did not contain the cross-shaped inner section. Instead two vertical guide rails in the center allowed a more flexible adjustment of the object carrier.

For tests, in which pacemaker and lead should be positioned on one layer, they were placed on a Plexiglas board with a fine, regular pattern of holes. Those holes firstly served as markers to detect imaging artifacts caused by the implants, secondly allowed

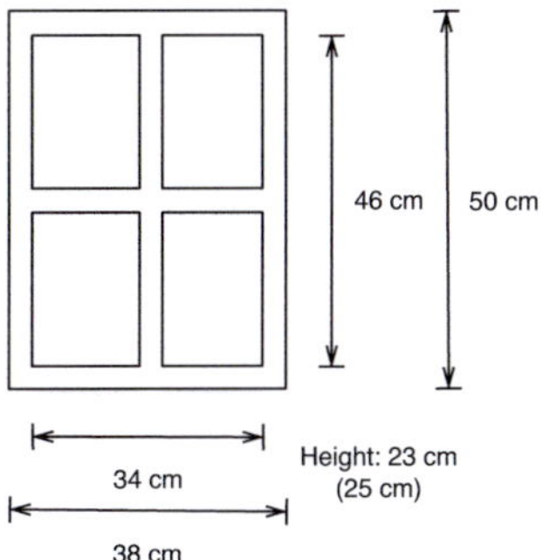

Figure 4.19.: Measurement phantom II: derived from phantom shown in figure 4.18.

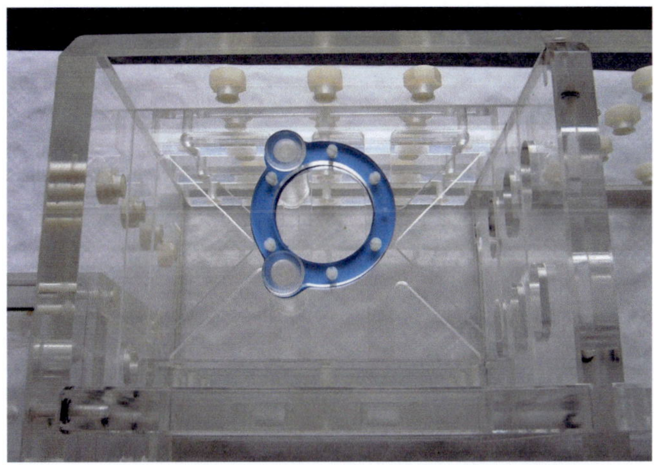

Figure 4.20.: Electrode and temperature probe carrier place in phantom.

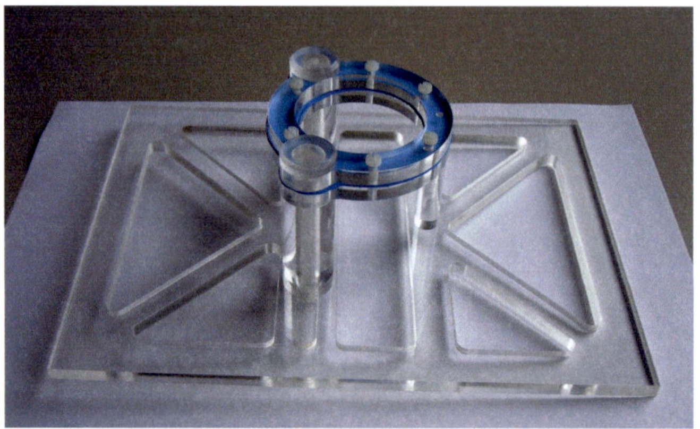

(a) Perspective view.

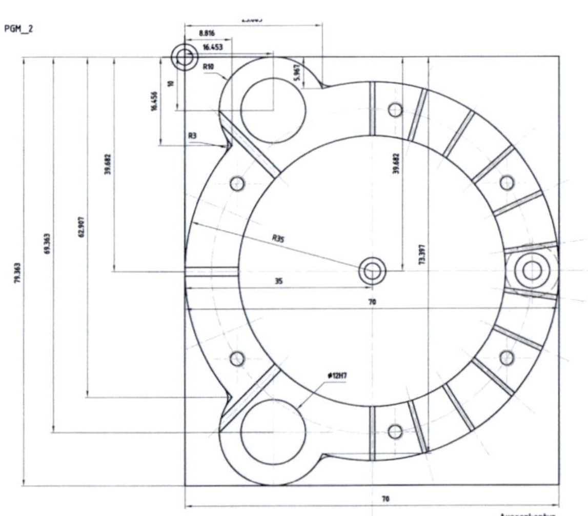

(b) Construction drawing.

Figure 4.21.: Lead and temperature probe carrier.

the use of strings to fix the devices and to reduce the blocking effect for the saline.

During the experiments, all phantoms were aligned to the center of the patient table. The table's axial position in the MRI device depended on the placement of the device under test inside the phantom. The aim was always to place the device in the field of view to achieve a maximum exposition to the RF field of the body coils.

4.2.4. Phantom Filling Materials

To ensure a milieu inside the phantom that was equivalent to the one inside of the human body, the phantom had to be filled with a liquid or gel. This filling, also called *Tissue Simulating Liquid* (TSL) or *Tissue Simulant*, affected several aspects of the experiment. It determined the wavelength of EM waves when propagating inside the TSL, the reflection at the boundaries, the attenuation of EM fields and the dissipation of potential heat caused by the tested objects. Besides impacts inside the phantom, it also acted as the main load for the RF coil, either birdcage or open MRI.

The composition of such a liquid or gel aimed at adjusting the dielectric parameters conductivity and permittivity as well as heat capacity and heat transfer. The first is relatively easy to control by adding NaCl until the required value of σ_r is reached, the second requires much more advanced ingredients. As a consequence, in most publications only the conductivity was controlled.

Inside a pure NaCl solution, heat is dissipated faster than inside human tissue. A more realistic milieu could be created by adding a gelling agent for example based on cellulose. It congeals the filling and therefore significantly reduces convective heat transfer.

Like the phantom itself, the ASTM standard also contains recommendations for fillings of the phantom with at least 30 kg. The filling material should have a conductivity of 0.4 to 0.8 S/m and a dielectric constant between 60 and 100 for measurements at 64 MHz and between 0.2 and 0.4 S/m for experiments at 1 kHz. The thermal parameters are specified with a heat capacity of 4184 J/kg °C. As an example, the standard references a formulation given by Konings et al. [110]. The resulting material has a conductivity of 0.25 S/m which is only suitable for 1 kHz measurements. Additional formulations for trials at 64 MHz were adopted by the authors of the standard from Chou et al. for brain

	General [110]	Brain[111]	Muscle [111]
Destilled water	n.a.	93%	91.48%
TX-151 Gelling Agent	–	7%	8.4%
(Oil Center Research)			
Polyacrilic acid	5.85 g/L	–	–
NaCl	0.8 g/L	0	0.12%
Conductivity [S/m]	0.25	0.52	0.8
ε_r	n.a.	78.1	79.8

Table 4.2.: Tissue simulating liquids as describd in ASTM Standard 2182-02a.

and muscle simulating liquids (see table 4.2) [111].

In the presented experiments, the phantoms were always filled with isotonic NaCl solution (0.9%) by Braun. In the literature, there is a wide range of conductivity settings described. Bassen et al. used 0.6% NaCl ($\sigma = 1.0\,S/m$) [21].

4.2.5. Pacemaker Models

As laid out in section 3.7 there are many different types of cardiac pacemakers. At the same time, profound knowledge about the detailed interior circuits is scarce. The only available parameter that is accessible to measurements is the input impedance.

One model was picked and used throughout all in-vitro tests: a *Frontier II* system by St. Jude Medical. It provided three connectors for sensing and stimulation leads. The connectors were encapsulated in a *Hysol epoxy* block and attached to the circuitry inside a titanium housing (see figure 4.22) [112]. The unused connectors must be sealed with a silicone plug to prevent saline or gel from causing shortcuts or forming connections between saline and electronic circuitries. On one sample device, the housing or can was opened to allow the measurements of interesting impedance parameters:

- Tip electrode/housing

- Tip electrode/ring electrode

- Ring electrode/housing

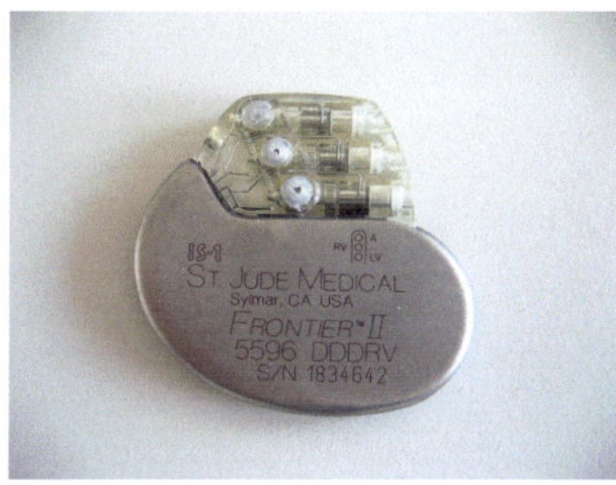

Figure 4.22.: Pacemaker device used for in-vitro test (*Frontier II* by St. Jude Medical).

Figure 4.23.: Inside view of pacemaker device used for in-vitro test (*Frontier II* by St. Jude Medical).

The measurements were conducted with an impedance analyzer (*4192A LF Impedance Analyzer* by Hewlett Packard). The connection to the fragile helix was achieved with a custom made adapter inserted into the core of the helix (see figure 4.24). The resulting values of the complex impedance ($\underline{Z} = R + jX$) were consecutively used for a linear fit. This fit served as the basis for an extrapolation of the values to higher frequencies like 42 and 64 MHz since the impedance analyzer only worked up to 13 MHz.

The values at 64 MHz were needed to parametrize a simulation with a pacemaker model and a connected lead (see section 4.1.6). Since the housing and lead consisted of PEC (Perfect Electric Conductor), the idea was to place lumped elements connecting the two. With it, the non-galvanic coupling of the housing and the lead should be imitated.

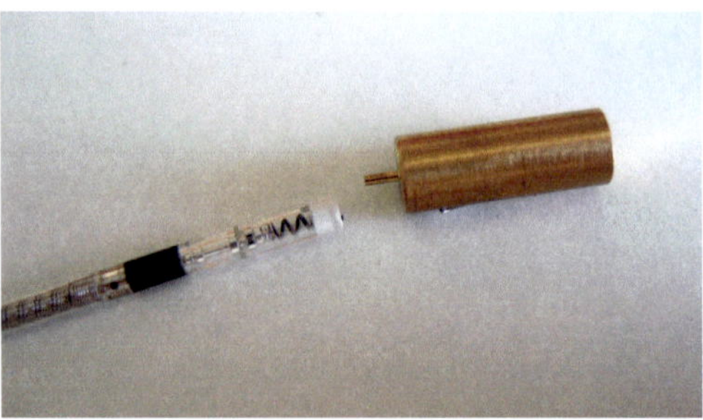

Figure 4.24.: Custom made adapter to connect pacemaker lead to impedance analyzer.

4.2.6. Pacemaker Leads

With the number of possible pacemaker positions, lead lengths, paths and orientations in mind, it was necessary to reduce the number of incorporated lead types that would be used for the in-vitro experiments. Available in three different lengths (45, 58 and 65 cm) was the bipolar *Capsurefix Novus* system by Medtronic (see figure 4.25). It can be fixed to the myocardium with an extractable helix that also serves as one electrode. The second electrode is implemented as a ring placed a few millimeters from the tip (see figure 4.25b). The insulation consists of silicone rubber, the metal part of *MP35N* (see section 3.6 for details of dielectric parameters).

The path of the lead inside the phantom followed commonly used implantation procedures (see section 3.8) with the pacemaker placed either right or left in the clavicle region. The lead was led from this point to the center and then towards the lower end of the phantom. In another configuration the lead was guided from the pacemaker site to the other side of the phantom to maximize the exposure to the changing magnetic field. In both cases, the remaining part of the lead was coiled around the pacemaker housing. To maintain a proper fixation while moving the phantom into and out of the MRi device, the lead and the pacemaker were fixed with thin rubber bands or nylon strings on a plate punctured with a regular raster of 1 mm thin equidistant (1 cm) holes.

4.2.7. Comparison of Effects of Open and Bore Hole MRI Systems

The experiments made to prove the theory of reduced induced currents in the open MRI system were done with the following configuration. The lead connected to the pace-

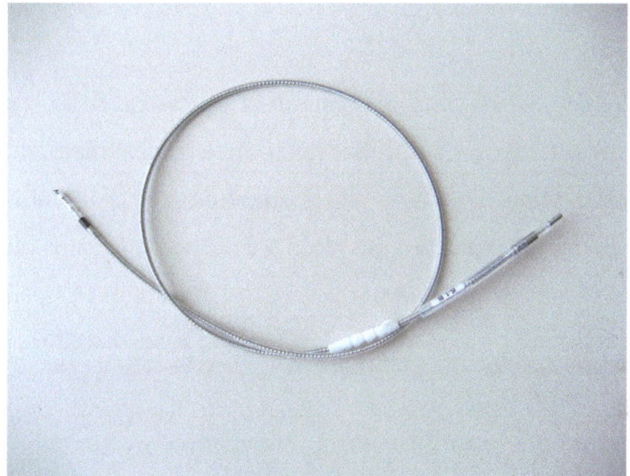

(a) Curled complete lead.

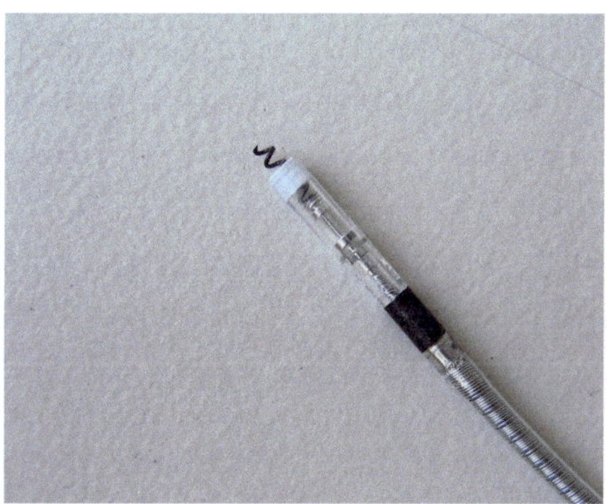

(b) Tip region.

Figure 4.25.: *Capsurefix Novus* cardiac pacemaker lead by Medtronic.

maker was placed and fixed on the object carrier forming an arch (see figure 4.26d). The setup of the experiments developed for the comparison of the bore hole and open MRI device is shown for an inclination of the pacemaker and lead of $45°$ in figure 4.26. The construction allowed a continuously adjustable inclination and vertical position of the object carrier. Three different angles were tested in both MRI types: $0°$, $45°$ and $90°$, all vertically centered.

The phantom was filled with about 39 l of saline to ensure full coverage also for the vertical positions and 4 temperature probes positioned as described above. The expected outcome was that in the bore-hole device the temperature rises would be reduced and in the open MRI device increased when the plate was tilted in the upright position.

4.2.8. Measurements with Thermocouple Elements

When trying to capture the most interesting information during an in-vitro experiment, the temperature change, the correct placement of the temperature probes relative to the tip of the lead is always difficult and prone to failures. Additionally the fiber-optic cables are rather stiff, must not be sharply bended and therefor require a considerable space around the lead's tip. After a lead has been implanted and fixed to the myocardium, it is basically impossible to capture the temperature without destroying surrounding tissue and tampering the original environment.

A potential workaround for this problem could be the inclusion of a temperature probe directly into the tip of the lead with a connection guided inside the lumen. Such a lead was custom made by *Osypka* for this project. It consisted of a thermocouple element bonded to the tip of the electrode and a very thin extension wire inside the lead till the distal end (see figure 4.27). There it was accessible again and could be further extended or connected to a measurement control unit.

In principle, a thermocouple consists of two different metals that are bonded together. It produces a voltage when exposed to temperature changes. When the change of the voltage is related to the temperature change, the thermocouple can serve as a temperature measurement system. Several metal combinations are commonly used today, each with a specific temperature range, stability, cost etc.

For the presented work, a K-type system was chosen that provided a sensitivity of $41\,\mu V/°C$. Type K sensors consist of chromel-alumel alloys (NiCrNi). The transfor-

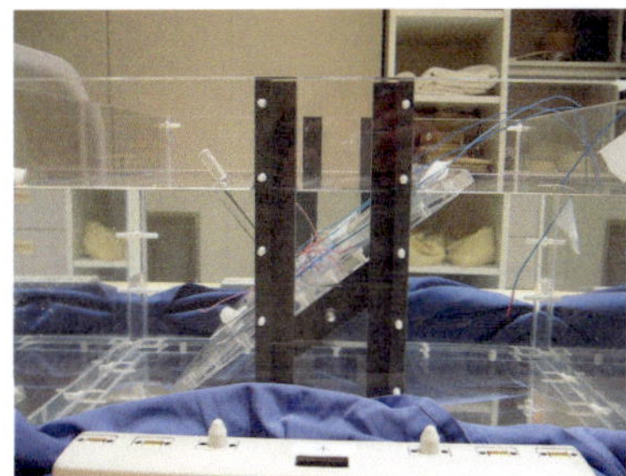

(a) Side view of phantom with object carrier and reference temperature probe in the background.

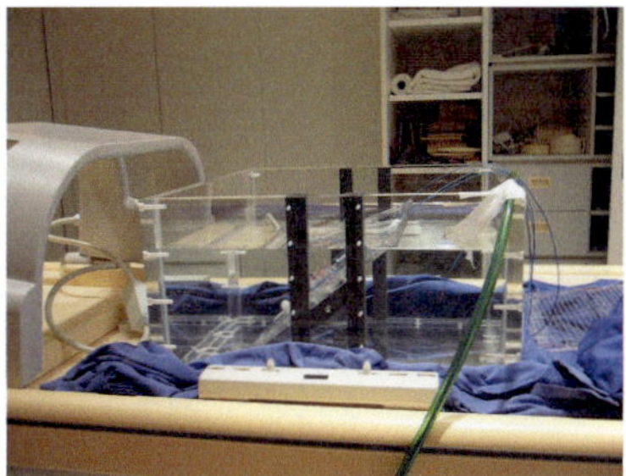

(b) Side view with open receive body coil.

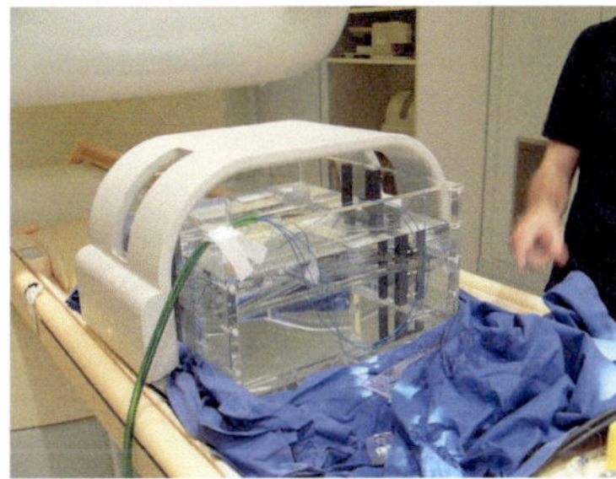

(c) Rear view with closed receive body coil.

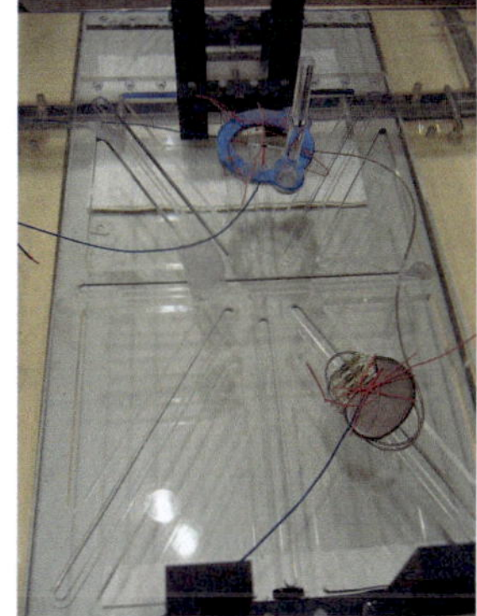

(d) Top view on the pacemaker and lead with temperature probes.

Figure 4.26.: Experimental setup showing phantom with inserted object carrier (inclination: 45°) placed on patient table of open MRI device. Two temperature probes are fixed around the tip, one at the pacemaker device and one in the distance for reference.

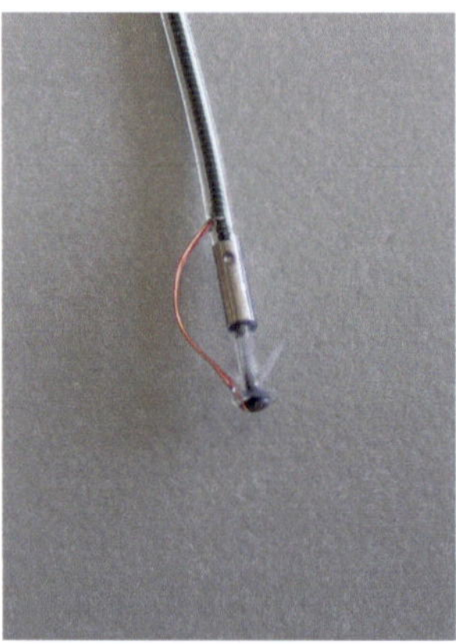

Figure 4.27.: Tip of lead with included thermocouple element and supply line.

mation into a temperature signal was processed by a model *5110* device (PeakTech). It offered an absolute accuracy of 0.5 % and detected temperature changes of 0.1 °C. An additional 4-channel device (*T-395*, PCE Group Europe) allowed the simultaneous observation of further measurement spots.

The biggest disadvantage of this measurement technique was its susceptibility for RF-related distortions. Since the feed lines were completely unshielded, the one meter or even longer cable acted as a perfect antenna for severe noise-related artifacts during MRI procedures.

5. Results and Discussion

The results presented in this chapter will be devided in the two main fields *Computer Simulations* and *In-vitro Experiments*. The structure of the sections is similar to the one of chapter 4 for better comparison.

5.1. Computer Simulations

Most questions covered in this work were studied by means of computer simulations. They allowed a fast and reproducible way to determine and analyze the occurring electromagnetic field distributions and additionally compute the resulting temperature changes for selected configurations. Basically the interesting aspect was always the energy deposited during exposure to the RF field. Since this is easily accessible via the SAR, for the majority of the cases, the evaluation was based on this parameter.

5.1.1. Validation of Huygens Box Method

The validation of the Huygens box was done initially for every simulation setup where this feature was used. In the following part, the similarity of the EM field distributions achieved with this method and with full field simulations will be compared based on the B-fields. They were extracted in all three spatial directions.

The presented result covers the birdcage coil and the ASTM-based Plexiglas phantom. Because the objects were later placed a few centimeters off the center of the coil, the values for the validation in the y- and z-direction were also extracted $10\,\text{cm}$ away from the center. The volume, where the objects to be tested should be placed, was $x = -0.03\,\text{m}$ to $-0.15\,\text{m}$, $y = -0.10\,\text{m}$ to $0.10\,\text{m}$ and $z = -0.12\,\text{m}$ to $0.12\,\text{m}$. As shown in figure 5.1, in all cases the B-field showed a sufficiently equal distribution in all axes. Only in the y-direction, a small deviation was observed. This distortion was of minor concern, because all objects were placed at $y = 0$ level where the fields matched each other.

An additional aspect in all simulations using the Huygens box was, whether the objects

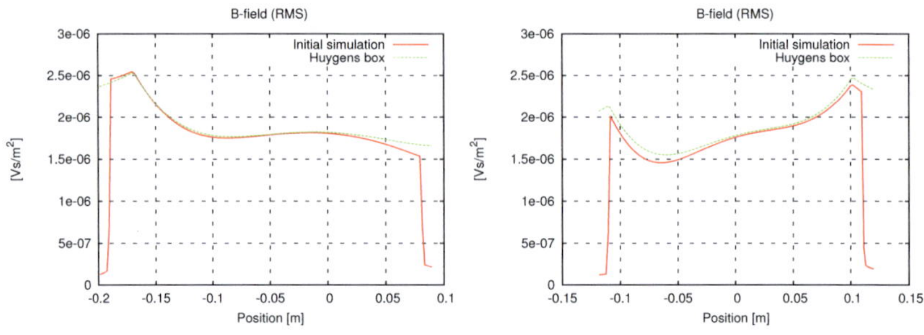

(a) B-field (RMS) along x-axis, y and z centered. (b) B-field (RMS) along y-axis, z centered, x = $-10\,$cm.

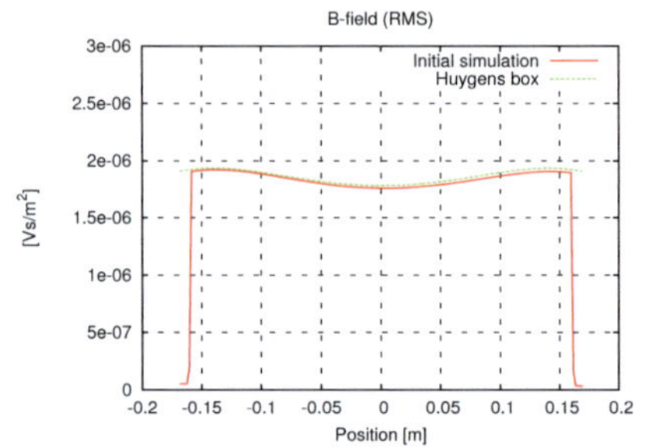

(c) B-field (RMS) along z-axis, y centered, x = $-10\,$cm

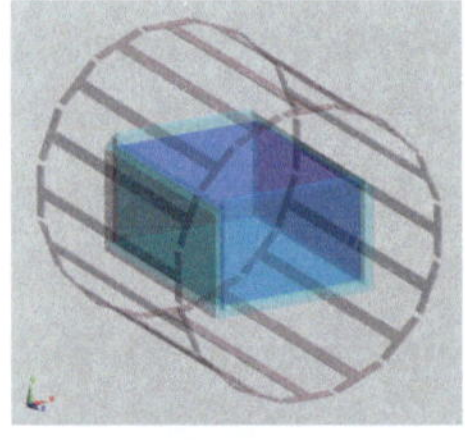

(d) Simulation setup.

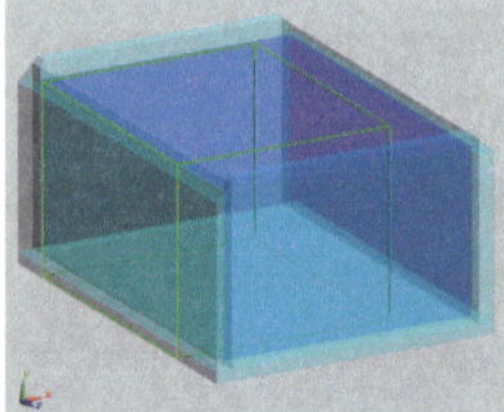

(e) Sensor inside Plexiglas box.

Figure 5.1.: B-field distribution in the birdcage coil loaded with plexiglas phantom, comparison between fields of whole field simulations and Huygens box.

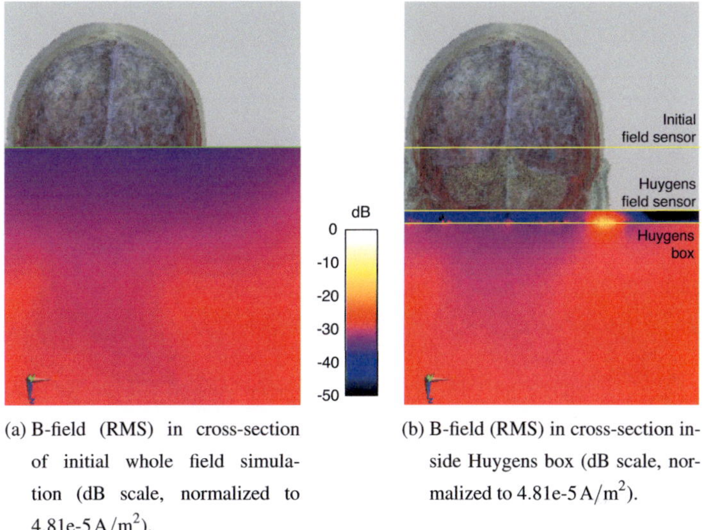

(a) B-field (RMS) in cross-section of initial whole field simulation (dB scale, normalized to 4.81e-5 A/m^2).

(b) B-field (RMS) in cross-section inside Huygens box (dB scale, normalized to 4.81e-5 A/m^2).

Figure 5.2.: Comparison between whole field simulations and Huygens box, B-field distribution in the birdcage coil loaded with the anatomical voxel dataset.

inside the box would interfere with the boundaries. By backscattering EM waves or effects on cutting-planes the TF/SF conditions could be harmed. Figure 5.2 illustrates this effect with an example located at the left ear region of the voxel dataset. Because of the ear's twisted structure and it's location directly at the boundaries, the B-field was back-scattered there and the area subsequently showed a superelevated field distribution.

This effect was also present in the E-field distribution and even affected a larger area (see figure 5.3). The whole free space above the shoulder showed a totally different field geometry compared to the initial simulation (see figure 5.3a).

Because the Huygens box only aimed at maintaining the field distributions inside its boundaries, any area monitored outside the box contained artifacts as illustrated by the prominently different regions in the upper part of figure 5.2b and 5.3b.

The similarity could also be shown for a setup with a metal object present. As illustrated in figure 5.4, the SAR distribution was equal for a whole field simulation and a Huygens box based one. The mesh resolution in the vicinity of the wire was equal (0.2 mm) in both cases and only varied in greater distance.

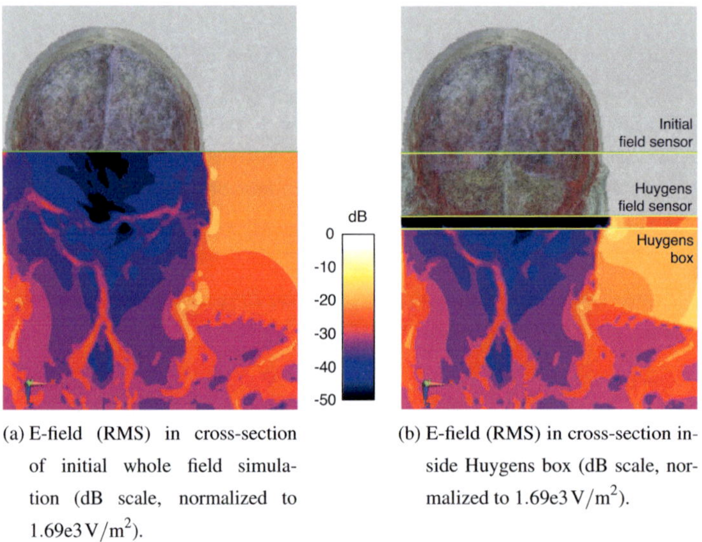

(a) E-field (RMS) in cross-section of initial whole field simulation (dB scale, normalized to 1.69e3 V/m^2).

(b) E-field (RMS) in cross-section inside Huygens box (dB scale, normalized to 1.69e3 V/m^2).

Figure 5.3.: Comparison between whole field simulations and Huygens box, E-field distribution in the birdcage coil loaded with the anatomical voxel dataset.

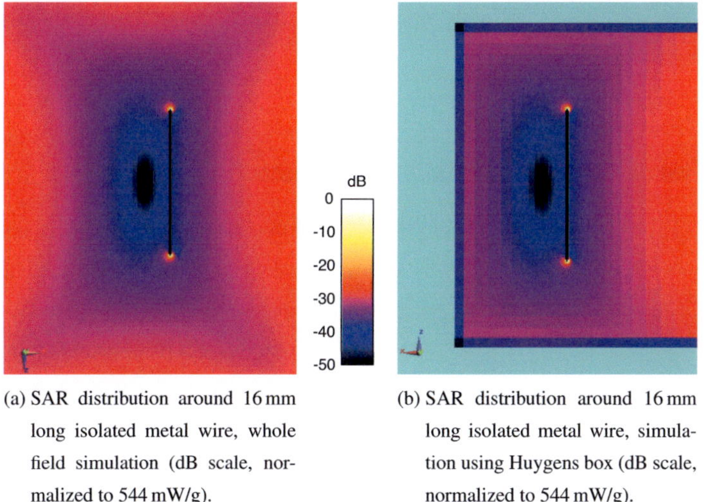

(a) SAR distribution around 16 mm long isolated metal wire, whole field simulation (dB scale, normalized to 544 mW/g).

(b) SAR distribution around 16 mm long isolated metal wire, simulation using Huygens box (dB scale, normalized to 544 mW/g).

Figure 5.4.: Comparison between whole field simulations and Huygens box, SAR distribution in the birdcage coil loaded with Plexiglas phantom.

Conclusion

The comparison of the whole, continuous field simulations showed that the Huygens box feature in SEMCAD produced appropriate results. For the tested configurations, the electromagnetic fields inside the Huygens box were quasi identical. Nevertheless the use of this feature required great care. If scattering objects inside the box were placed too close to the boundaries, the conditions of the Total-Field / Scattered-Field approach were broken. A second aspect was scattering at the cutting plane of structures that got truncated by the boundaries of the box. For objects like the Plexiglas phantom, it was no problem because the outer form of the phantom matched the Huygens box' boundaries. The situation was a little bit different for simulations with the voxel model. Due to the irregular and curved surface, the reflections of the RF fields showed complex patterns that reached the boundaries and induced disturbances there.

As a consequence, the voxel model required larger dimensions of the Huygens box. If the box did not cut the voxels and provided enough surrounding free space (padding) to let the backscattered EM waves fully decay, the disturbances could be avoided. Even then, the simulation times could be drastically reduced or more complex configurations assessed respectively. The simulation of the electrode following the vessels would otherwise have been impossible.

5.1.2. Validation of Perfect Electric Conductor

The results of the comparison between a wire set to realistic conductivity values ($\sigma_r = 0.968e6\,S/m$) and the same object set to PEC showed a maximum difference in the SAR values of 31 mW/g. In the first case, the maximum SAR was 544 mW/g, in the second 513 mW/g – a difference of 6%. Since the PEC configuration resulted in higher values, simulations with these settings produced useful results with even a little safety margin. When evaluating the SAR and current density maps (see figure 5.5 and 5.6), the SAR distributions proved to be sufficiently equal. Small deviations were caused by slightly different meshes but the corresponding values could often be found just in the neighboring layers. The current densities decreased towards the center of the wire and the values here were lower in the MP35N case than in the PEC case (see figure 5.8). The last aspect could be explained with the conductivity assigned to both objects. The PEC was attributed with a quasi indefinite conductivity while the one of MP35N was only $\sigma_r = 0.968e\text{-}6\,S/m$.

Additionally, the two simulations were evaluated using the SAR histogram method for all values inside the Huygens box. As illustrated in figure 5.7, a comparison of the SAR distribution of the wire modeled as PEC and the second with MP35N properties showed the same strong identity as seen before with the plots. Only minor outliers were present, while the majority of the curves matched each other.

Conclusion

The results of the comparison of a metal wire modeled as PEC and as MP35N proved the validity of the PEC formulation in SEMCAD. Since it led to a reduced simulation time and produced field distributions that were even a little bit too high, the results could be regarded as reasonable accurate.

5.1.3. Anatomical Voxel Model and Wire

When simulating EM field distribution, the anatomical datasets still represent the golden standard. Assuming that the dataset used is highly detailed and properly segemented,

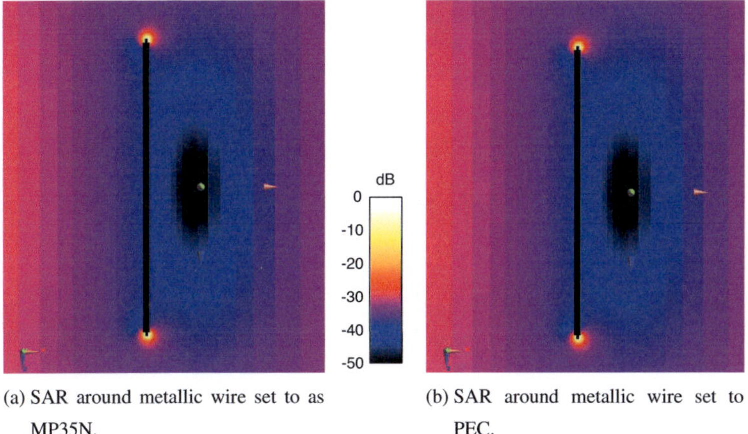

(a) SAR around metallic wire set to as MP35N.

(b) SAR around metallic wire set to PEC.

Figure 5.5.: Comparison of MP35N and PEC parametrization of metal objects. SAR distribution around 16 cm long metallic wire, isolated with silicone rubber, 2 mm bare ends, placed in birdcage coil (dB scale, normalized to 544 mW/g).

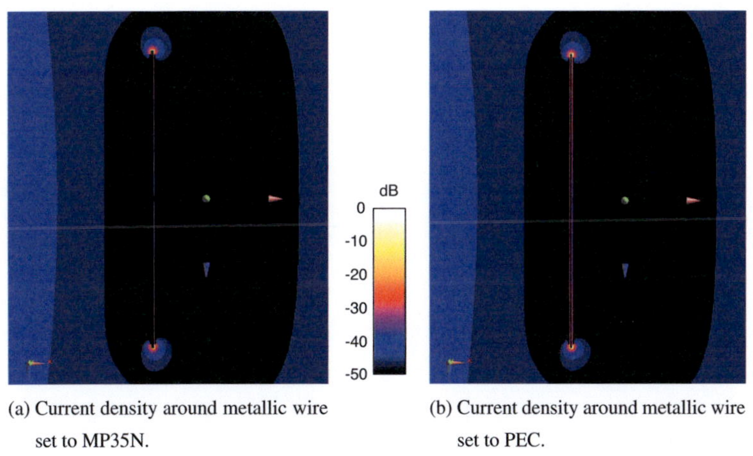

(a) Current density around metallic wire set to MP35N.

(b) Current density around metallic wire set to PEC.

Figure 5.6.: Comparison of MP35N and PEC parametrization of metal objects. Current density distribution around 16 cm long metallic wire, isolated with silicone rubber, 2 mm bare ends, placed in birdcage coil (dB scale, normalized to 2.7e3 A/m^2).

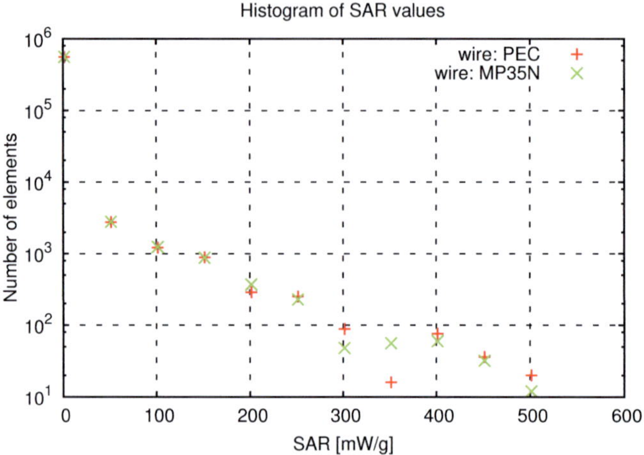

Figure 5.7.: Histogram of SAR values around 16 cm long metallic wire, isolated with silicone rubber, 2 mm bare ends, placed in birdcage coil.

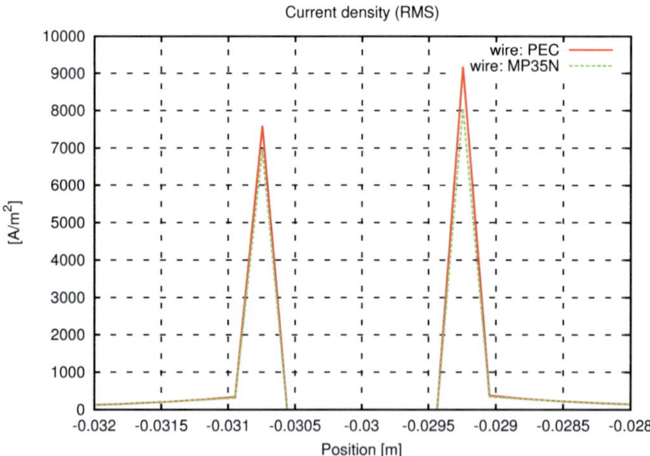

Figure 5.8.: Comparison of current density values, extracted at a line crossing the tip of a 16 cm long metallic wire, isolated with silicone rubber, 2 mm bare ends, placed in birdcage coil.

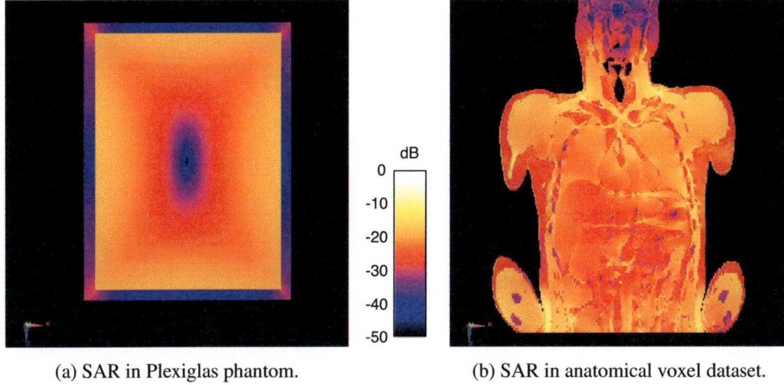

(a) SAR in Plexiglas phantom.

(b) SAR in anatomical voxel dataset.

Figure 5.9.: SAR distributions inside Plexiglas phantom and anatomical voxel dataset in birdcage MRI coil (dB scale, normalized to 0.0979 mW/g, input power adjusted to 1 W).

the EM wave propagation is more realistic than in a homogeneous phantom. At the the same time, this introduces a significant amount of complexity to the simulation. Hence the computational cost, time and memory requirements can increase tremendously. As described above, the voxel dataset required a larger Huygens box than the Plexiglas phantom.

First, the influence of the inner volume with its tissue type boundaries is shown with a SAR plot (see figure 5.9b). Due to the different dielectric properties of the various tissues, the EM waves could not cross the volume constantly but were attenuated and/or reflected. The normalization to 1 W was done to assure that in both cases the birdcage coil was fed with the same amount of power regardless of the load present. The second simulation contained a wire that was placed inside the voxel model in a subcutaneous pectoral position (see figure 5.10).

5.1.4. 3D Surface Models

The first results acquired were the point clouds as output from the 3D scanner (see figure 5.11). With two measurement campaigns, a total number of 9 people were registered while standing and additionally when sitting. The intention was that deforming an up-

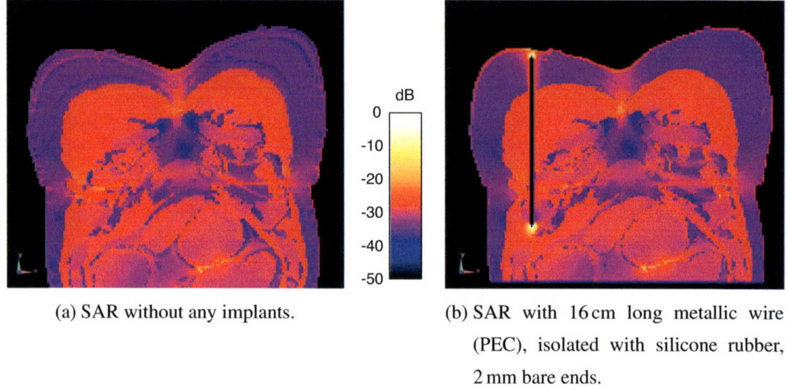

(a) SAR without any implants.

(b) SAR with 16 cm long metallic wire (PEC), isolated with silicone rubber, 2 mm bare ends.

Figure 5.10.: SAR distribution inside anatomical voxel model, placed in open MRI coil (dB scale, normalized to 300 mW/g).

right model into a sitting position would always be less precisely than capturing this posture directly.

After de-noising and post-processing, each dataset was available as a polygon model (see figure 5.12). To illustrate the occurring artifacts, two were left in figure 5.12b: the closed hole on the inside of the right arm and the hair.

5.1.5. Generic Organ Models

This section will illustrate the application of newly developed deformation rules for the creation of generic organ models. Examining the results of the isotropic scaling showed that this method induced incorrect proportions in the models. As visible in the example of a 5 year old child in figure 5.13, the isotropic model was too high and not wide enough [113, 114].

Conclusion

The results achieved with the orthogonal transformation procedure showed good accordance to real datasets extracted from the *Visible Family* (see section 4.1.3). Based on this work, a set of organs can now be quickly generated by scaling the reference organs to the required age level. Though, not all scaling factors from the mathematical phantoms were fully reliable. Further research is recommended to ensure a constant quality for all

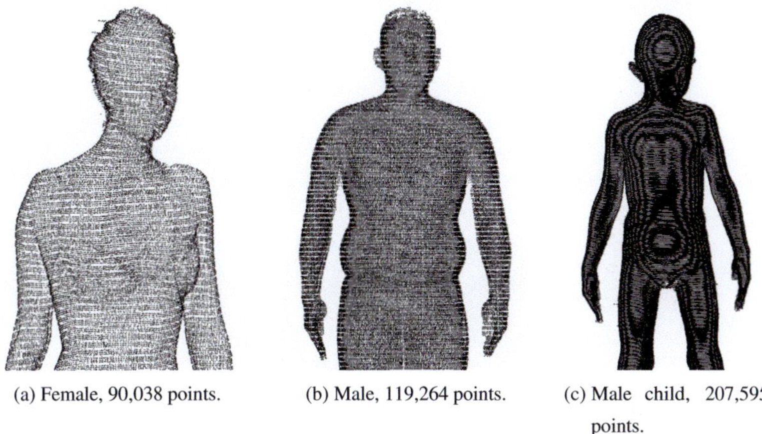

(a) Female, 90,038 points.

(b) Male, 119,264 points.

(c) Male child, 207,595 points.

Figure 5.11.: Point clouds acquired with 3D scanner.

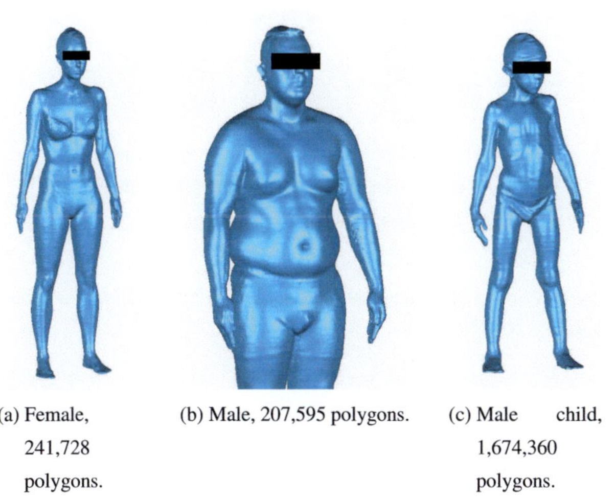

(a) Female, 241,728 polygons.

(b) Male, 207,595 polygons.

(c) Male child, 1,674,360 polygons.

Figure 5.12.: Polygon surface models derived from the point clouds shown in figure 5.11.

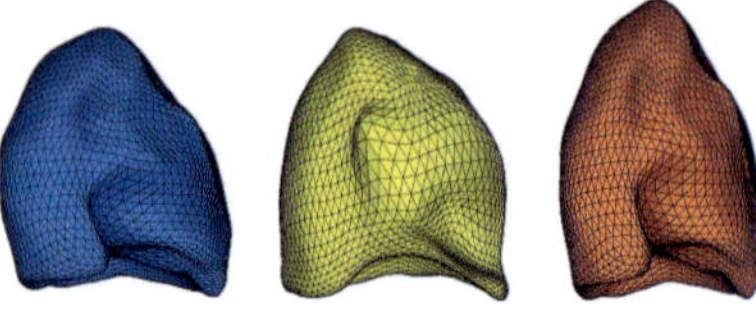

(a) Comparison of a orthogonally transformed model, a real dataset and a isotropically scaled model (front view).

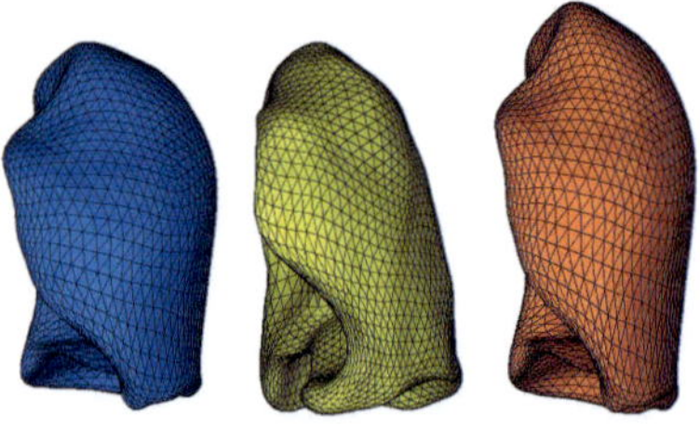

(b) Comparison of a orthogonally transformed model, a real dataset and a isotropically scaled model (side view).

Figure 5.13.: Comparison of generic lung models generated with different transformation methods, age 5 years.

incorporated models.

Not covered by this work was the subsequent placement of the organs in the surface models. Here as well, in a follow-up project the necessary rules could be developed.

5.1.6. CAD Models

Open MRI Coil

The newly developed *RF coil* for the open MRI was validated by comparing the computed magnetic fields with the already existing birdcage coil. For all three spatial dimensions, the extracted B-field (RMS) is plotted in figures 5.14 and 5.15. To examine the stability of the B-field, in all diagrams two cases are displayed: for the empty coil and when loaded with the plexiglas phantom.

In all three directions, the open MRI coil generated a stable, homogeneous B-field spanning the whole (FOV) region. The intensity was in the same order of magnitude (mikro tesla) even though a little lower than in the birdcage coil. While the open MRI coil provided a field of about 1.4 to $1.6\,\mu\mathrm{T}$ (see figure 5.15), in the birdcage's FOV region values of about $2\,\mu\mathrm{T}$ were observed (see figure 5.14).

Influence of Position inside Phantom

The influence of the tested object's position on the genesis of induced currents was examined by a simulation study placing a wire in 5 different positions off the center of the phantom and the coil respectively (see figure 5.16b for the setup). Because the variation of B_1 over time ($\mathrm{d}B_1/\mathrm{d}t$) increased when approaching the inner border of the coil, the induced currents and as a consequence the SAR should increase as well. As shown in figure 5.16a, the assumption could be proved. To take all voxels into account, again the histogram method was chosen to illustrate the results. But not only the overall distribution changed, also the maximum values significantly rose: at $-3\,\mathrm{cm}$ it was $544.23\,\mathrm{mW/g}$, at $-7\,\mathrm{cm}$ $2{,}950.8\,\mathrm{mW/g}$ and at $-11\,\mathrm{cm}$ finally $6{,}894.59\,\mathrm{mW/g}$.

Conclusion

As expected, the simulations showed a strong correlation between the position of the wire and the observed SAR values. This had also been found by Mattei [30], who reported similar findings from an in-vitro study.

Although the results seem to hint at an easy method to reduce the risk for the patient, it is

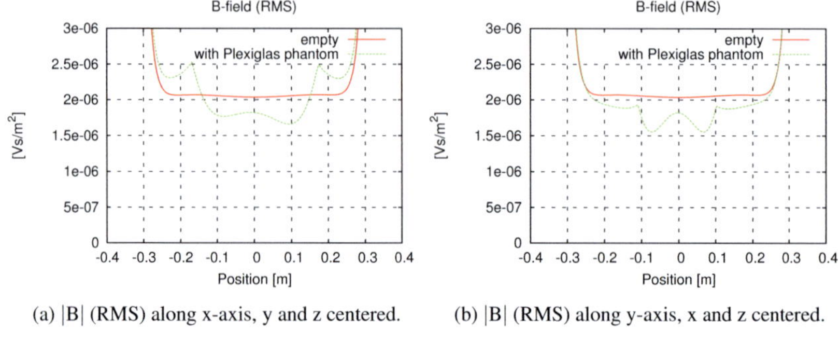

(a) |B| (RMS) along x-axis, y and z centered.   (b) |B| (RMS) along y-axis, x and z centered.

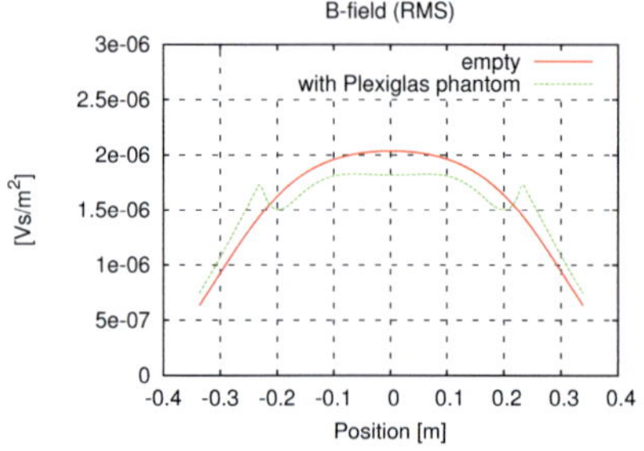

(c) |B| (RMS) along z-axis, y and y centered.

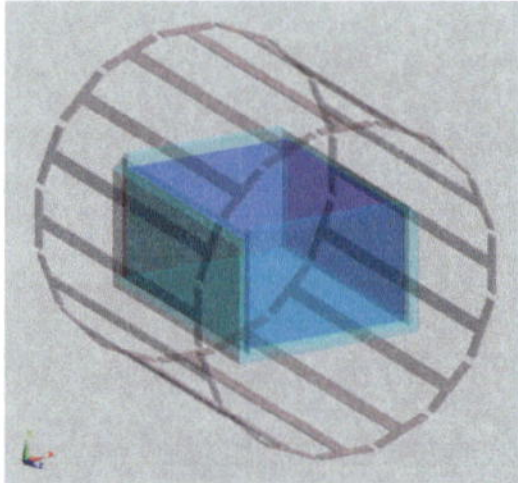

(d) Simulation setup.

Figure 5.14.: Distribution of |B| in the birdcage MRI coil, empty and loaded with a phantom along, all three axes.

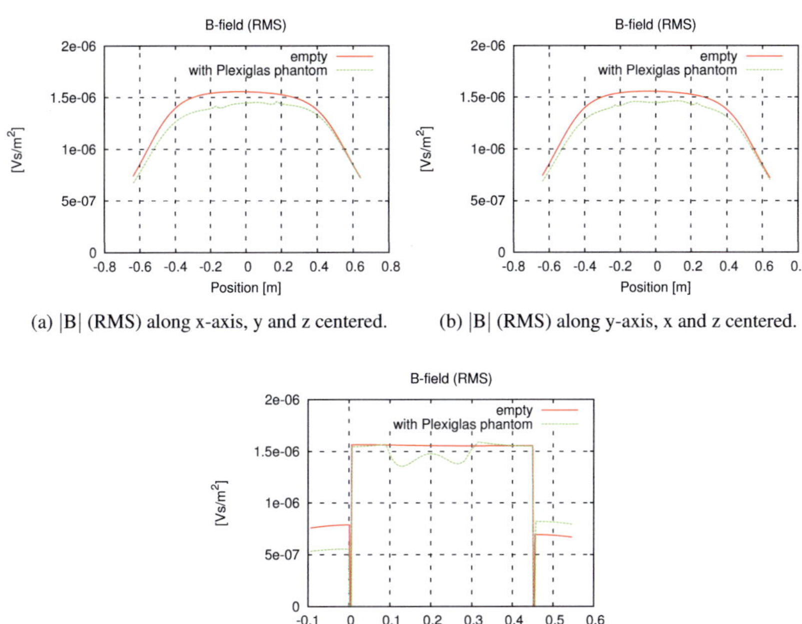

(a) |B| (RMS) along x-axis, y and z centered.

(b) |B| (RMS) along y-axis, x and z centered.

(c) |B| (RMS) along z-axis, x and y centered.

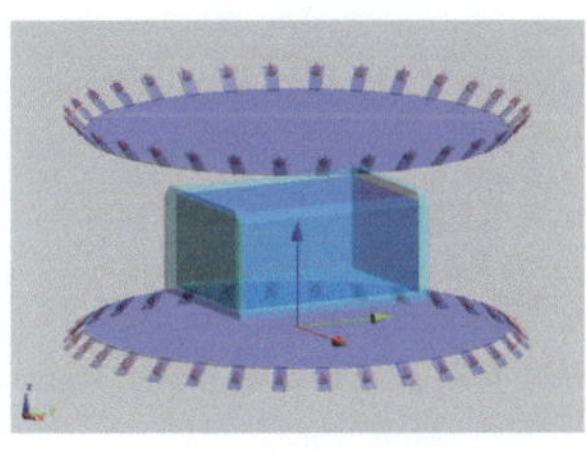

(d) Simulation setup.

Figure 5.15.: Distribution of |B| in the open MRI coil, empty and loaded with a phantom, along all three axes.

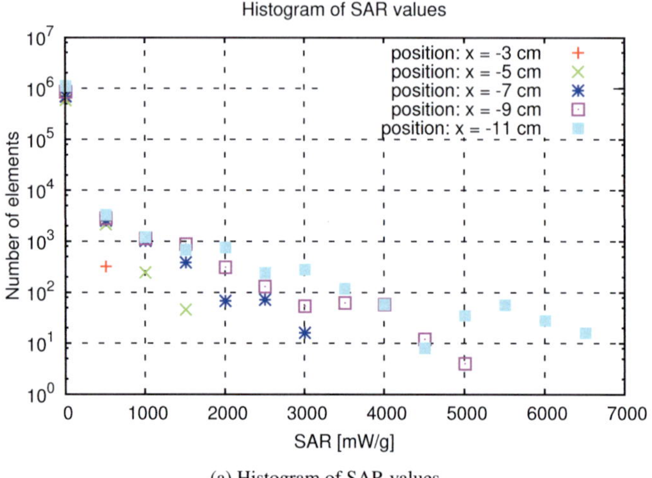

(a) Histogram of SAR values.

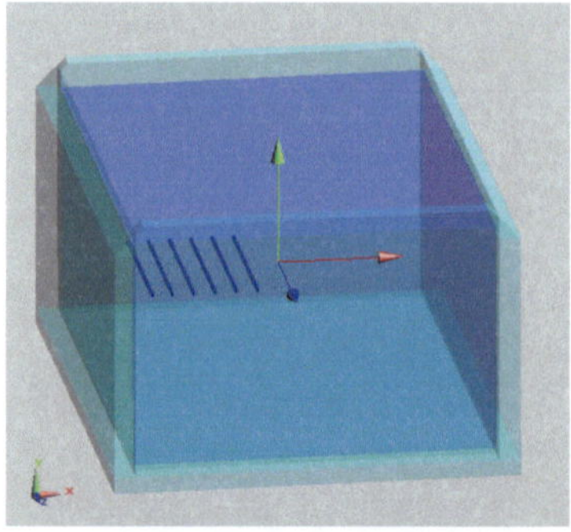

(b) Positions of the metal wires inside the
plexiglas phantom.

Figure 5.16.: Histogram of SAR values around 16 cm long metallic wire, isolated with silicone rubber, 2 mm bare ends, placed in birdcage coil at x = −3, −5, −7, −9, −11 cm.

not very useful in daily practice in bore-hole MRI devices. Normally, the position of the patient can only be reasonably adjusted in the z-direction. Although in an open MRI the patient could also be moved in the lateral position, the field orientation of the RF signals is different. As shown later, this already causes a much more prominent reduction in the SAR by itself.

Unconnected Pacemaker Lead

The contribution of a complex modeling of the pacemaker leads construction was assessed with an unconnected lead equipped with a helix tip (see figure 5.17c) [115]. The same model was also used to illustrate the influence of the SAR avaeraging volume. As observable in figures 5.17a and 5.17b, the helix structure was clearly visible in the SAR distribution. The highest values could be found in the surroundings of the ring electrode. The fine structures required a mesh step in the helix region of 0.1 mm, which again was only feasible due to the Huygens box feature.

Conclusion

The results achieved with the complex lead model together with the fine mesh provided valuable insights on the effects close to such a miniature structure. Although one could have expected the highest values close to the fine structure, they were located at the ring electrode. Furthermore, the inside of the helix was way less affected than the outside. This is positive, because therein heat would accumulate even better due to the enclosed, very low perfused tissue. In contrast, the increased heating around the ring electrodes took place in the well perfused lumen of the heart where deposited heat would be quickly deported by the constant blood flow. The same effects were found in the in-vitro experiments (see section 5.2.2). Another reason for high resolution models and meshes is the skin effect, due to a concentration of the currents on the outer surfaces of the conducting wires.

When evaluating the maximum SAR values, they were considerably lower than for the wires discussed in the previous section. At the same position of $-7\,$cm, the wires had induced an SAR of up to 2,905 mW/g while the complex model provoked only 461.7 mW/g. This could have been caused by the helix structure inside the supply line acting as an inductor and dampening the RF waves.

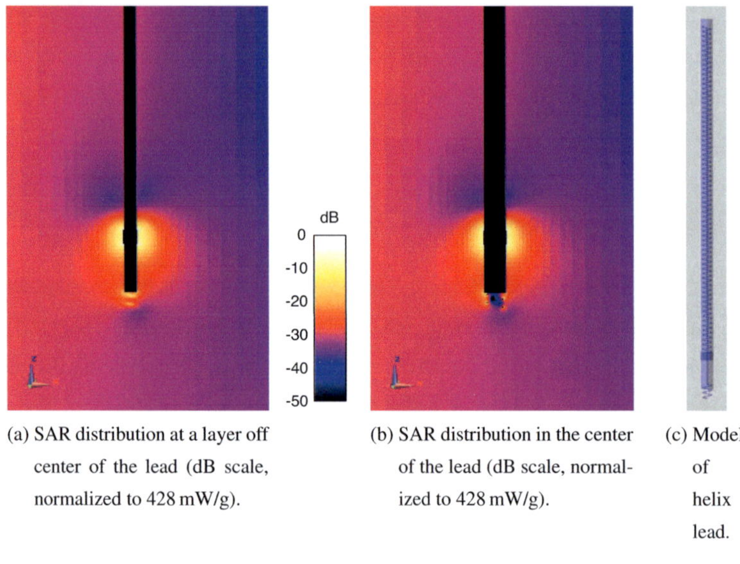

(a) SAR distribution at a layer off center of the lead (dB scale, normalized to 428 mW/g).

(b) SAR distribution in the center of the lead (dB scale, normalized to 428 mW/g).

(c) Model of helix lead.

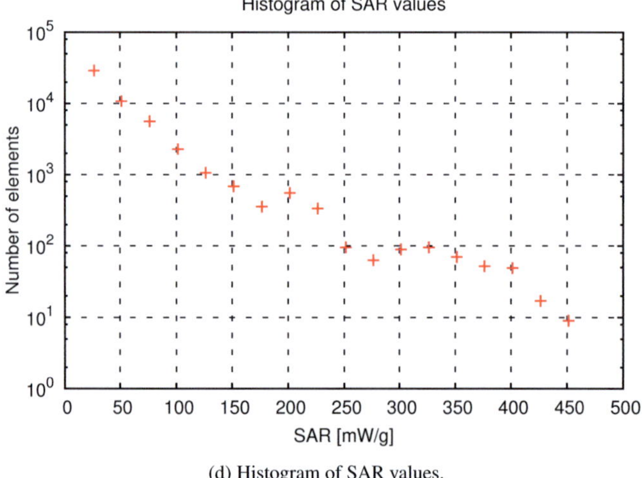

(d) Histogram of SAR values.

Figure 5.17.: SAR distribution around helix pacemaker lead, placed in birdcage coil at $x = -7$ cm, y and z centered.

Pacemaker Lead Following Anatomical Structure

In the following, the results acquired with a lead following the vessel system inside the anatomical voxel dataset will be presented. The lead was placed in the Plexiglas phantom, since the voxel dataset was too large for the available computer systems (see figures 5.18b and 5.18c). The details of the model have been described in section 4.1.6. The maximum SAR values observed were 2,609.63 mW/g located at the subcutaneous end of the lead and 376 mW/g at the opposite end (see figure 5.18a).

Conclusion

The mentioned maximum SAR values were a mixture of the values observed for the shifted wires. The subcutaneous end was located about -10.7 cm from the center, the other one at 3.7 cm. The results for the first (2,609.93 mW/g) could be related to the wire at -11 cm (maximum SAR: 6,894.59 mW/g) and the second (376 mW/g) to the wire at -3 cm (544.23 mW/g) regarding their position in the B_1 field. Although just being a singular result, it could be shown that a lead following a real-life path produced a different field distribution. A next step is to attach a pacemaker model as described in the next section. The *implantation* of a complex lead model in the same way still would fail due to the lack of enough computational power to process the resulting mesh.

Pacemaker Model with Lead

This section covers the computed field distributions for the pacemaker model connected to a straight wire model (see figure 4.10). Three different configurations for the connection of housing and lead were computed:

- single 5 nF capacitor

- combination of a 87 Ω resistor and a 1.782 µH inductor (low pass filter with cut-off frequency of 48.82 MHz)

- combination of a 87 Ω resistor and a 178.2 µH inductor (low pass filter with cut-off frequency of 0.49 MHz)

As illustrated in the histogram (see figure 5.20a), the setup with the capacitor showed the highest energy absorption in the surrounding liquid. The peak SAR value of 15,639.62 mW/g was about ten times higher than found in the setup with the smaller inductor (1,145.3 mW/g).

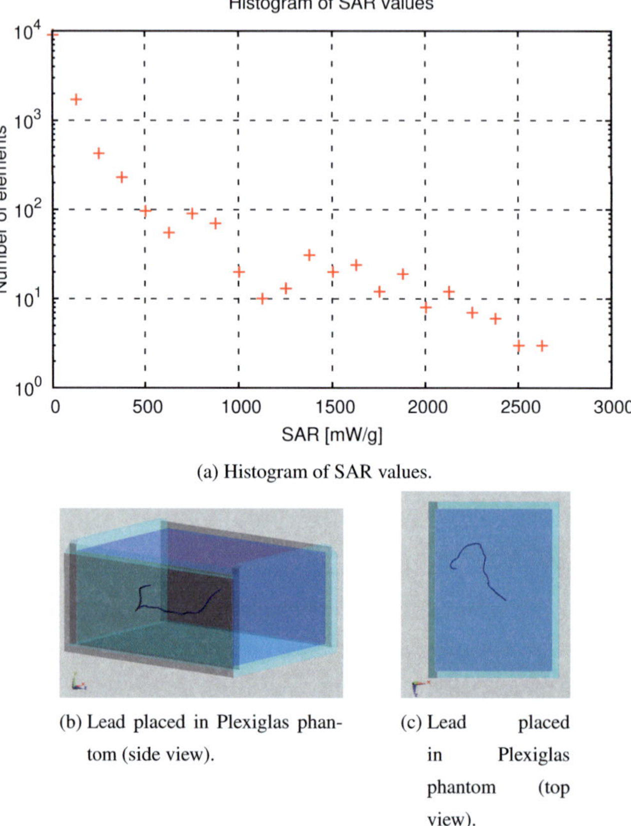

(a) Histogram of SAR values.

(b) Lead placed in Plexiglas phantom (side view).

(c) Lead placed in Plexiglas phantom (top view).

Figure 5.18.: SAR histogram and model of lead following anatomical structure, placed in Plexiglas phantom and birdcage coil.

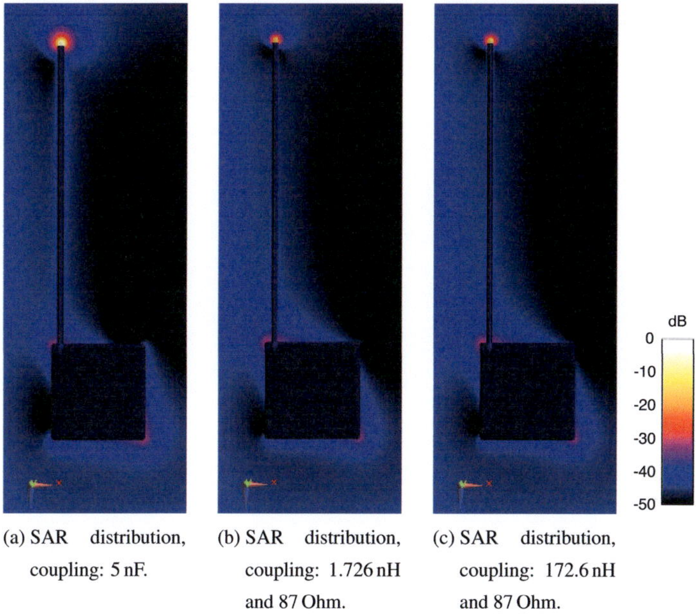

(a) SAR distribution, coupling: 5 nF.

(b) SAR distribution, coupling: 1.726 nH and 87 Ohm.

(c) SAR distribution, coupling: 172.6 nH and 87 Ohm.

Figure 5.19.: SAR distribution around model of pacemaker and lead, coupled with different lumped elements (dB scale, all normalized to 1.56e4 mW/g).

The third setup with the larger inductor produced a maximum SAR of 1,659.3 mW/g, the same order of magnitude as for the second configuration, although a little higher than for the smaller inductor.

Conclusion

The results of the pacemaker/lead model simulations indicated that a pure capacitive coupling of the lead and the housing would lead to significantly higher SAR values than a combination of inductor and resistor. Because a capacitor is reducing its impedance with increasing frequency ($\underline{Z}_{cap} = 1/j\omega C$), it actually connects the housing to the lead instead of decoupling it.

The idea of adding an inductor significantly reduced the observed SAR values. With a coiled pacemaker lead this might be even further improved. Increasing the size of the inductor by the factor of 100 did not introduce any remarkable additional benefit. The introduction of the inductor in general contributed the majority of the effect. This

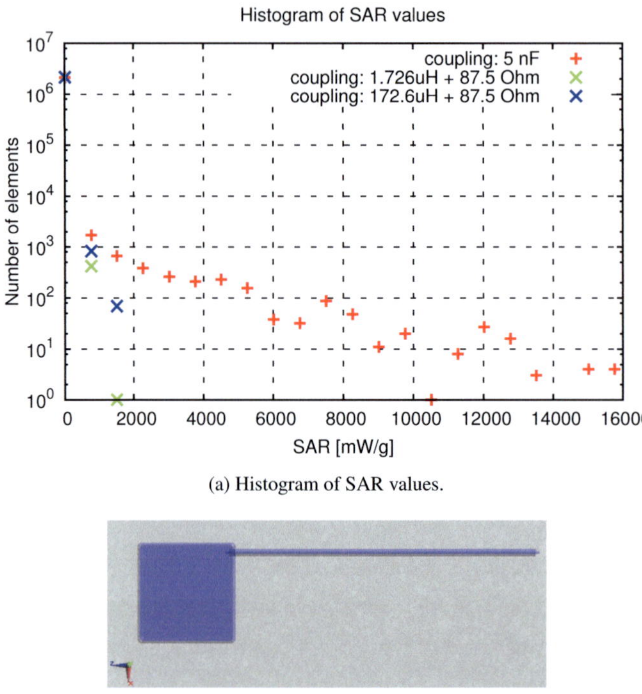

(a) Histogram of SAR values.

(b) Model of pacemaker and lead.

Figure 5.20.: Histogram of SAR values around model of pacemaker and lead, coupled with different lumped elements, placed in Plexiglas phantom and birdcage coil.

also implied that the extrapolated values used in the simulation may be an acceptable approximation.

The producers of recent pacemakers and leads have already included similar techniques in their products. They claimed to have modified the density of the windings along the lead to create small inductors that counteract the induction of currents.

5.1.7. Influence of Fiber-optic Temperature Probes on the Electromagnetic Field Distribution

The influence of the temperature probes introduced into the region of interest was negligible. As shown in figure 5.21c, the E-field was nearly unaffected by the fiber-optic object in the region close to the tip, only in the distal region, a small alteration was found. The SAR distribution in figure 5.21d confirmed this. Due to the basically non-conducting properties of the fiber, the computed SAR values inside the fiber (compare equation 3.34) were nearly equal to zero.

Conclusion

As a consequence, a fiber-optic temperature probe only captures heat coming from the outside. It did neither negatively affect the E-field distribution nor contribute any own heating. The use of the probe could therefore be regarded as reasonably safe.

5.1.8. Influence of Location of Fiber-optic Temperature Probe

The influence of the position of the temperature probes relative to the neuralgic part of the tested object like electrodes was determined with a temperature simulation. The heating was caused by a wire model placed in the Plexiglas phantom filled with saline. Three point sensors captured the heating during the simulations in a distance of 0, 1 and 2 mm distance from the bare tip of the wire in x- and z-direction. The setup was stimulated with a birdcage coil operating on a TSE sequence. As clearly visible in figures 5.22a and 5.22b, the captured temperature values decayed rapidly with an increasing distance. Already after 1 mm, the temperature measured was only 30% of the one at the surface of the wire. In 2 mm it further decreased to 10%. In a gel-filled phantom, the effects would be a little bit different. Because there the convection is reduced by the gelling agent, the induced heat is later visible at the probe but at the same time the following decay is less steep.

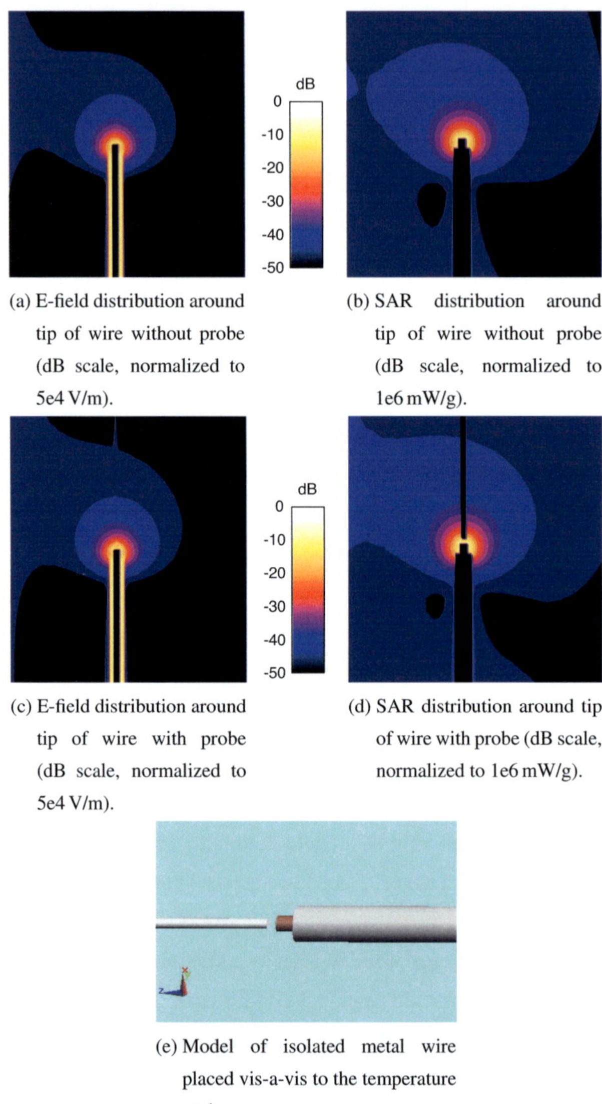

(a) E-field distribution around tip of wire without probe (dB scale, normalized to 5e4 V/m).

(b) SAR distribution around tip of wire without probe (dB scale, normalized to 1e6 mW/g).

(c) E-field distribution around tip of wire with probe (dB scale, normalized to 5e4 V/m).

(d) SAR distribution around tip of wire with probe (dB scale, normalized to 1e6 mW/g).

(e) Model of isolated metal wire placed vis-a-vis to the temperature probe.

Figure 5.21.: Influence of temperature probe on the EM field distribution [116].

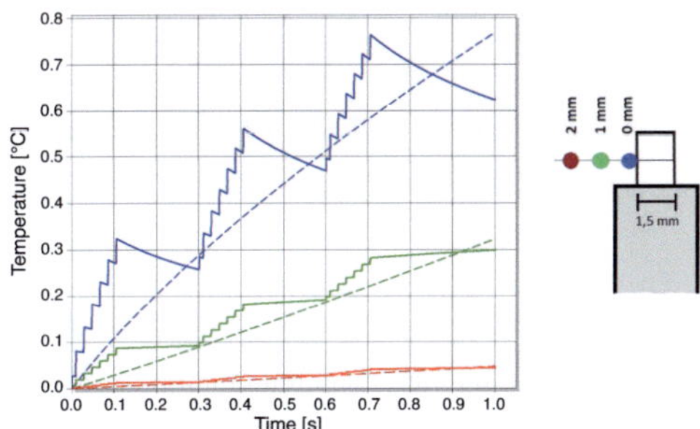

(a) Computed temperature change caused by TSE sequence.

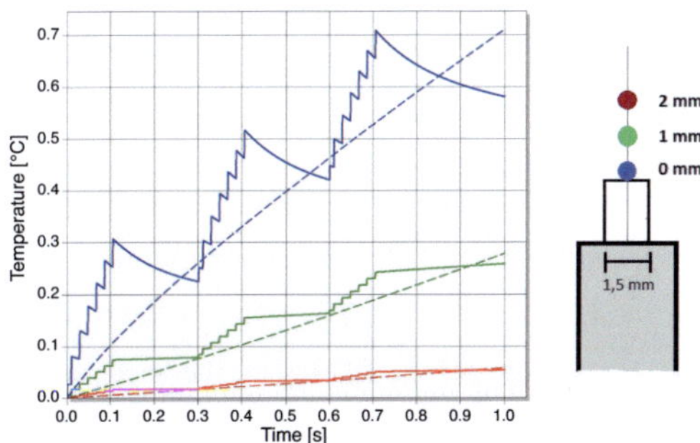

(b) Computed temperature change caused by TSE sequence.

Figure 5.22.: Influence of temperature probe distance on the captured temperature [116].

Conclusion

The results of the temperature simulations described above have a strong impact on in-vitro measurements and also on plans to do in-vivo experiments. In contrast to in-vitro tests, here the tip of the electrode is very difficult to access with fiber-optic probes. Since the temperature decayed so rapidly with increasing distance from the leads, it is crucial in experiments to assure a proper positioning of the probes in direct vicinity of the examined object. Already a distance of 1 mm would show an untruly low temperature distribution. This could lead to an underestimation of the situation and the hazardous potential of the examined setup. The findings here matched the results of Neufeld et al. [32]. They also had found a significant correlation between location of the temperature probe and measurement error.

5.1.9. Influence of Phantom Filling Parameters on Temperature Distribution

This section covers the temperature simulations focusing on the influence of the phantom filling. As shown in figure 5.23a, the temperature elevation was higher when the phantom was filled with TSL. When the wire was placed in the NaCl solution, less heating was observed. After switching off the RF signal, the accumulated thermal energy quickly dissipated into the surrounding volume with similar speeds. The histogram of the temperature values confirmed these findings: as illustrated in figure 5.23c, more voxels with higher temperatures values occurred in the TSL setup than in the NaCl filled phantom (see figure 5.23b).

Conclusion

The temperature simulations with varying filling parameters proved a difference for configurations between NaCl solution and TSL. After 2 s the deviation of the resulting temperature increase was approx. 8.3 %. This discrepancy should be considered when performing in-vitro measurements. However experiments with NaCl solution can produce reliable results if the observed temperature values are assessed accordingly.

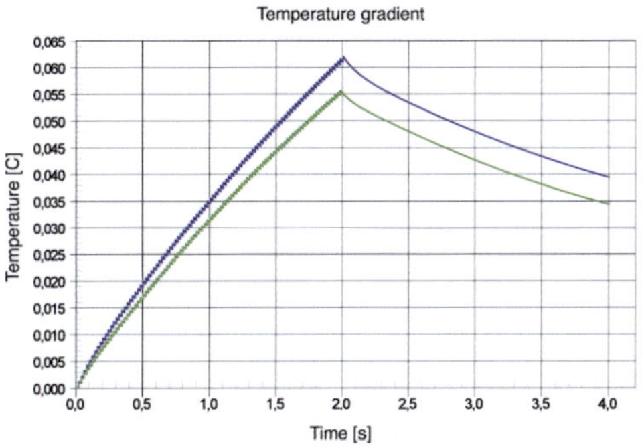

(a) Temperature gradients at the tip of the wire (blue = tissue simulating liquid, green = NaCl).

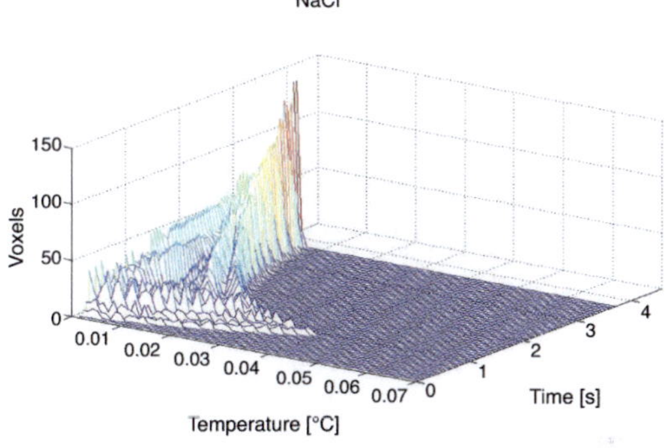

(b) Histogram of temperature values (NaCl).

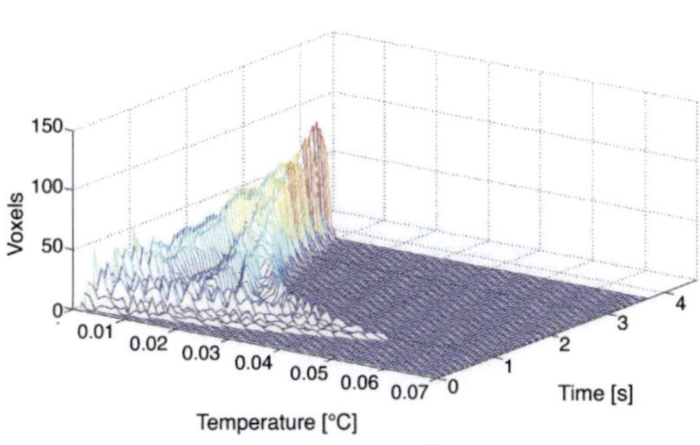

(c) Histogram of temperature values (tissue simulating liquid).

Figure 5.23.: Temperature distribution around metal wire, placed in Plexiglas phantom filled with NaCl or tissue simulating liquid, birdcage coil [116].

5.1.10. Realistic MRI Sequences versus Scaled Continuous Wave Excitation

The temperature simulations based on the implementation of the saturation recovery sequence gave the following insights. Already in the small breaks after an RF pulse the heat at the tip quickly dissipated into the surrounding volume (see figure 5.24a). Furthermore, when the exposure stopped totally after 10 seconds the accumulated heat immediately declined (see figure 5.24b). Additionally, after 10 s the accumulated heat was considerably lower for a repetition times of 50 ms (approx. 0.5 °C) than for the shorter 10 ms one (approx. 2.7 °C) [117].

The comparison of the full-featured sequence and the scaled continuous wave stimulation (see section 4.1.9) showed that the simplification was a valid approach. At the very beginning of the stimulation a small difference could be found (see figure 5.25a). Later both temperature gradients matched each other properly (see figure 5.25b).

Conclusion

The temperature simulations gave valuable results not only for further simulations but also for in-vitro experiments. The scaled signal could approximate the effects caused by the pulsed excitation accurately. Therefore, a detailed modeling of MRI sequences might be skipped as long as the overall power is known to scale the continuous signal. Even small breaks of only a few seconds could reduce accumulated heat effectively and already a little longer delay between the RF pulses lowered the overall temperature considerably. Regarding implant heating in general, the results showed that excitations with a reduced amplitude but larger duty cycles induce the same amount of currents and consecutively the same amount of heating.

5.1.11. Encapsulation of Pacing Lead

The encapsulation of the tip with a medium that offered the same dielectric properties as fibrotic tissue reduced the maximum SAR by a factor of more than two. Although the layer was just 1.25 mm thick, the SAR decreased from 1,150 mW/g to 453 mW/g. In figure 5.26c, the SAR has been extracted for both cases along a line crossing the tip horizontally.

Conclusion

The implications of the results achieved with the encapsulated lead are difficult to define.

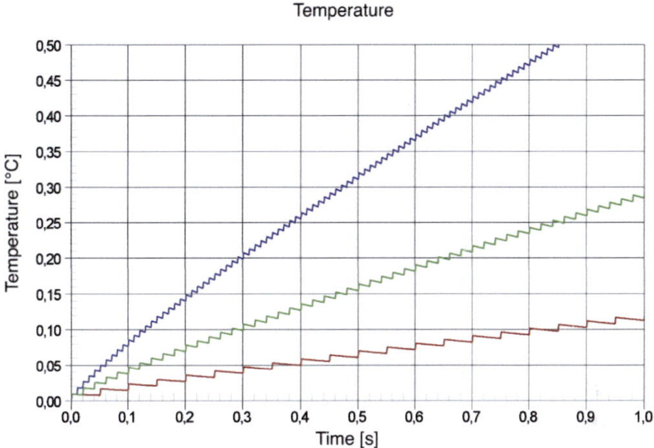

(a) Temperature for T_R=10 ms (blue), 20 ms (green), 50 ms (red).

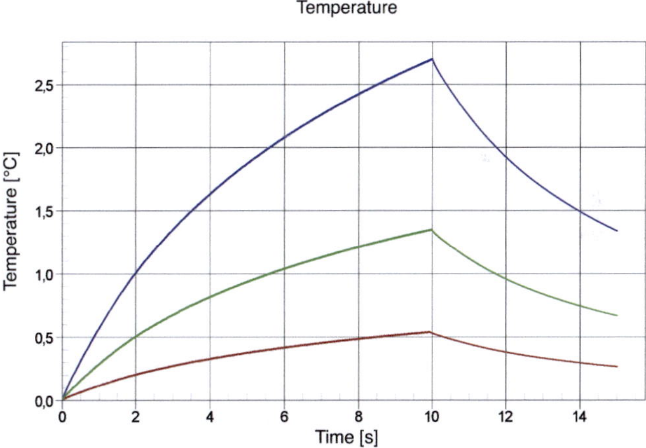

(b) Temperature for T_R=10 ms (blue), 20 ms (green), 50 ms (red).

Figure 5.24.: Temperature in the vicinity of an isolated straight wire, birdcage coil excited with saturation recovery sequence [116].

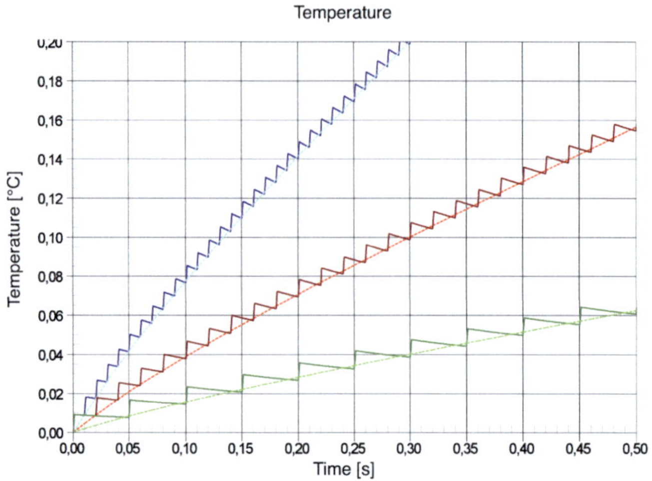

(a) Temperature for T_R=10 ms (blue), 20 ms (green), 50 ms (red).

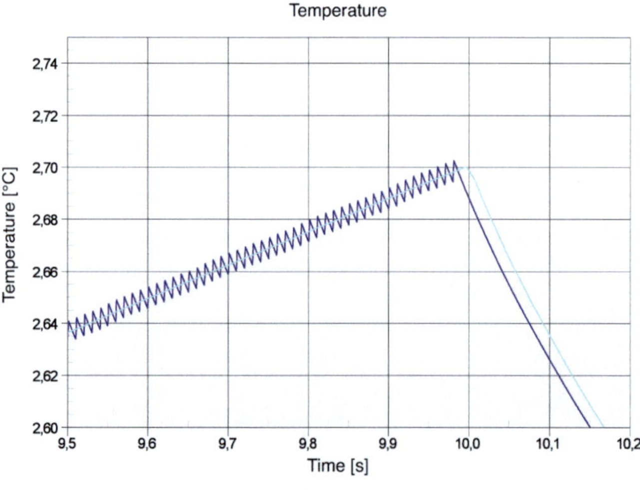

(b) Temperature for T_R=10 ms after 10 s.

Figure 5.25.: Comparison of temperature in the vicinity of an isolated straight wire, birdcage coil excited with saturation recovery sequence and scaled continuous wave [116].

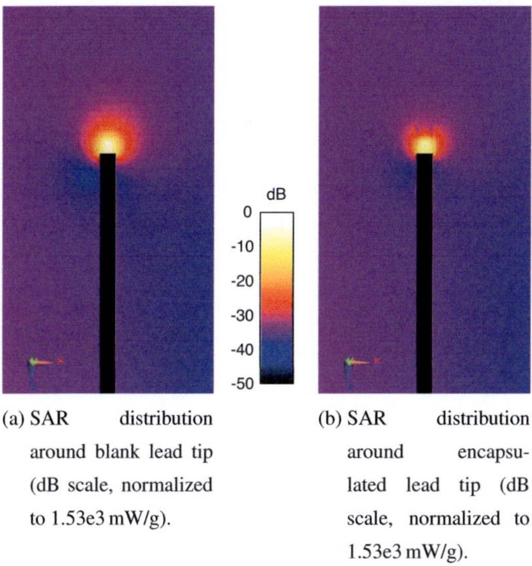

(a) SAR distribution around blank lead tip (dB scale, normalized to 1.53e3 mW/g).

(b) SAR distribution around encapsulated lead tip (dB scale, normalized to 1.53e3 mW/g).

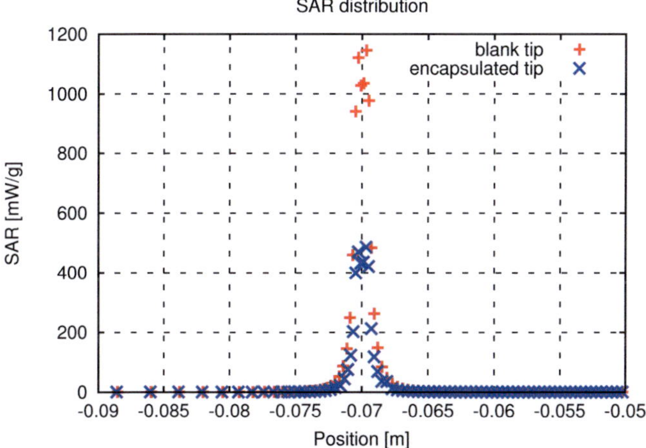

(c) SAR distribution along x-axis crossing the tip region of the lead, z centered.

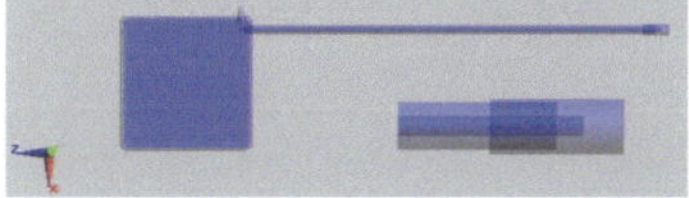

(d) Model of pacemaker and encapsulated lead tip.

Figure 5.26.: SAR distribution around model of pacemaker and encapsulated lead tip, placed in Plexiglas phantom and birdcage coil.

Obviously, even the thin additional low-conducting layer reduced the transit of the currents into the surrounding tissue. But as described in the literature overview, the genesis and intensity of the encapsulation can vary significantly between patients and over time [43]. Since the quantification of encapsulation is hard to detect for an implanted lead (all studies were made on explanted leads), it would be just speculative that the tip is coated for example after two years. Therefore, no reliable information for the physician could be derived from these simulation results. Nevertheless, Stoke et al. had found that at least in low perfused vessels an increased encapsulation could be expected [44].

But electrodes are not only encapsulated by fibrotic tissue. When fixing a lead in the atrium, the helix shaped tip is penetrating the myocardium and as a consequence buried in the heart's wall. In this case, the tip is also surrounded by a medium with a conductivity lower than that of blood (myocardium: σ=0.678 S/m, blood: σ=1.207 S/m at 64 MHz). Thus a similar effect could be expected. As a consequence, less energy would be deposited in the myocardium.

5.1.12. Comparison Between Open and Bore Hole MRI

The comparison of the results acquired with the classic birdcage and the newly developed open MRI coil was based on SAR values. As also shown in the in-vitro experiments (see section 5.2.3), effects like SAR and temperature changes inside the open MRI coil were generally lower than in the birdcage coil. Nevertheless, within one coil type, the results were based on the same system and are therefore very well comparable.

The first examined complex provides results computed with the birdcage model. Looking at figure 5.27, the effect of the inclination of the wire was very obvious. When orientated perpendicular to the rotating B_1-field, the wire caused the highest SAR values of about 8,468 mW/g. At 45° the value decreased a little to 5,209 mW/g. The most most prominent change occurred for the 90° position. Because the wire was now parallel to the rotation plane of the B_1-field, the maximum SAR dropped to 15.33 mW/g – about 0.19 % of the value at 0°.

In the case of the open MRI coil, the effect was less prominent. From 479.46 mW/g for an inclination of 90° the SAR only decreased to 9.26 mW/g (1.9 % when orientated parallel to B_1, which is the regular orientation in open MRI.

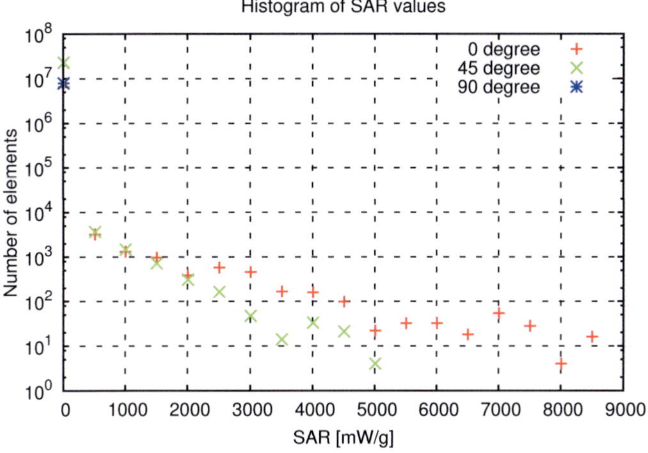

(a) Histogram of SAR values.

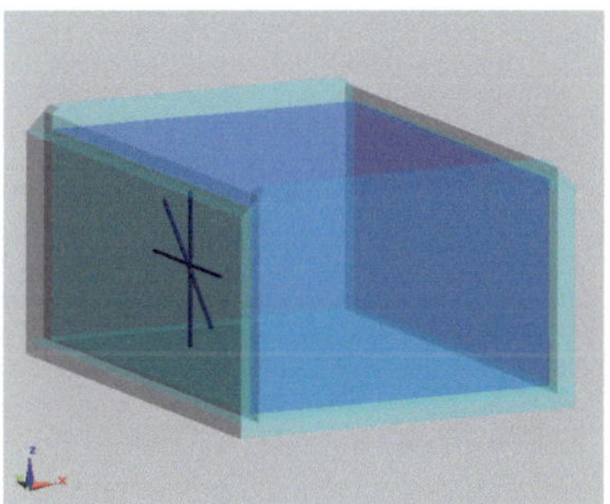

(b) Positions of the metal wires inside the plexiglas phantom.

Figure 5.27.: Histogram of SAR values around a 16 cm long metallic wire, isolated with silicone rubber, 2 mm bare ends, placed in birdcage coil. Illustrated are inclinations of $0°$, $45°$ and $90°$.

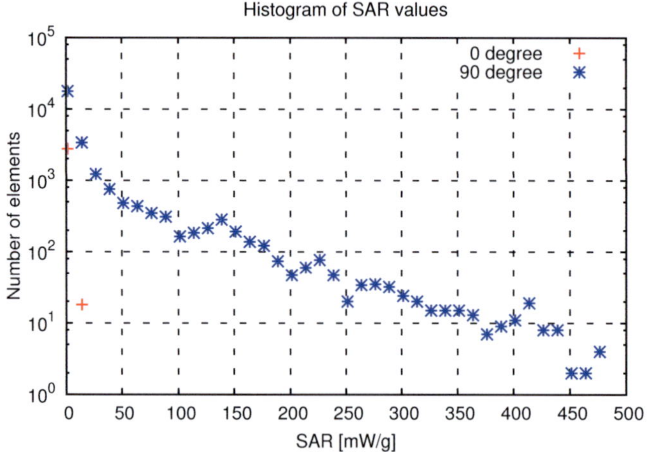

(a) Histogram of SAR values.

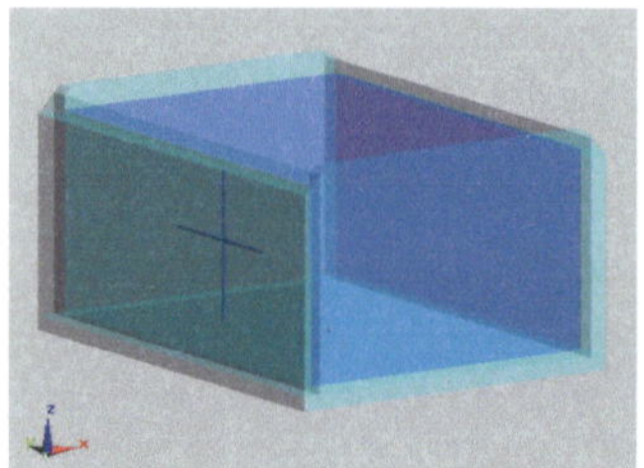

(b) Positions of the metal wires inside
the plexiglas phantom.

Figure 5.28.: SAR distribution around a 16 cm long metallic wire, isolated with silicone rubber, 2 mm bare
ends, placed in open MRI coil, illustrated are inclinations of $0°$ and $90°$.

Conclusion

The comparative study with the birdcage and the newly developed MRI coil could prove the hypothesis of an reduced risk of induced currents in open MRI. An orientation of the wire parallel to the plane of the rotating B_1-field reduced the maximum SAR to 0.19% in the birdcage and 1.9% in the open MRI case. Since in the open MRI coil the SAR maximum was already much lower (479 mW/g) than in the birdcage (8,468 mW/g), the less prominent reduction might be of less concern. Additionally the observed patterns furthermore confirmed the validity of the open MRI coil model.

5.1.13. Influence of SAR Averaging Volume

In the following, the impact of different averaging volumes on the SAR values will be described. In case of the fine helix structure, the computed SAR distributions showed a strong dependency on the averaging volume (see figure 5.29).

In the saline solution that surrounded the lead model, a mass of 1 g corresponded to a volume of $1\,cm^3$. The tip of the helix was 6 mm long, so the averaging volume was larger than each of the two electrodes. For 10 grams, the averaging cube had a side length of ca. 2.15 cm and therefore could include even both electrodes at one. Comparing the results for the not-averaged, the 1 g and the 10 g case identified a significant blurring of the SAR results. The peaks distributed around the helix were totally annihilated. Furthermore, with increasing averaging volume, the observed maximum value decreased from 462 mW/kg to as low as 2.69 mW/kg.

Conclusion

The results have clearly shown that, when using SAR distributions to identify small hot spots in the energy deposition, only not-averaged or averaged over 0.1 g SAR values could be regarded as useful. As also described by Mattei [31], the larger averaging volumes like 10 g in European and 1 g in US regulations will blur local heating. Even the 0.1 g Mattei proposed resulted in a 18 times smaller maximum SAR [31]. Since the regulations were specified for situations without implants, the above findings do not affect normalization when used for whole body SAR or local studies without implants.

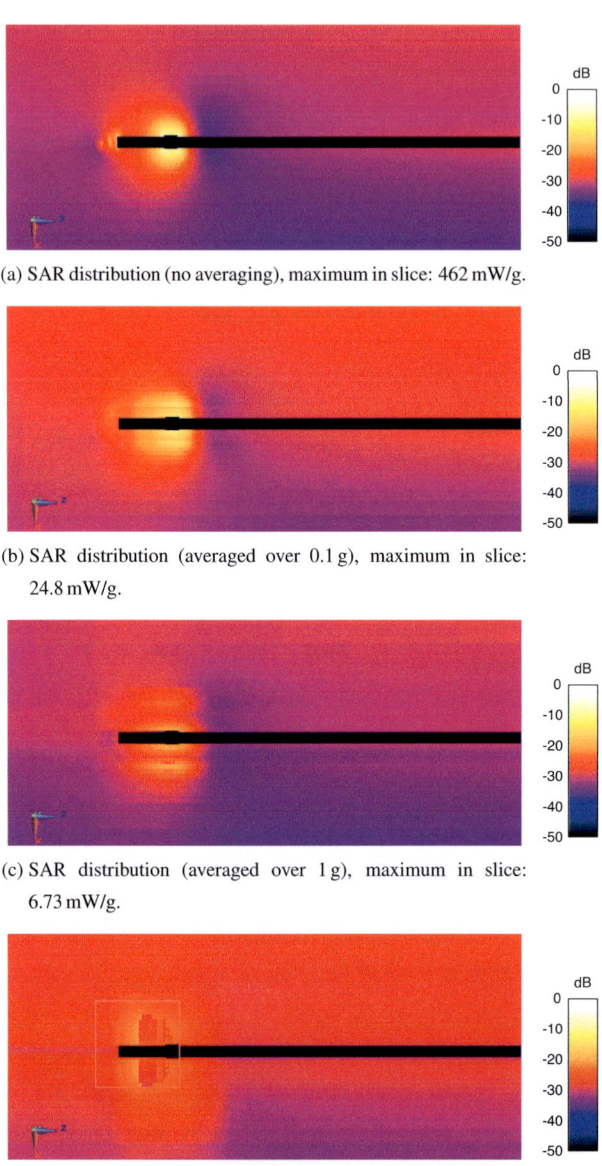

(a) SAR distribution (no averaging), maximum in slice: 462 mW/g.

(b) SAR distribution (averaged over 0.1 g), maximum in slice: 24.8 mW/g.

(c) SAR distribution (averaged over 1 g), maximum in slice: 6.73 mW/g.

(d) SAR distribution (averaged over 10 g), maximum in slice: 2.69 mW/g.

Figure 5.29.: Influence of averaging volume on SAR distribution around pacing lead with helix tip, placed in Plexiglas phantom (dB scale, all normalized to 462 mW/g).

5.2. In-vitro Experiments

Parallel to the computer simulations, selected configurations were examined in in-vitro experiments. The following section will cover all results achieved during those experiments.

5.2.1. Impedance Measurements of Pacemaker/Lead System

In this first section, the results of the input impedance measurements of the pacemaker/lead system will be presented. For lower frequencies (<1 kHz), the behavior was as described by Medtronic and Osypka. In the region of 800 Hz, the imaginary term prevailed ($-47,240\,\Omega$ vs. $510\,\Omega$, 45 cm lead) significantly and the real part could be neglected (see figures 5.30a and 5.31a). Regarding the system now as a capacitor, its capacity could be determined as

$$\frac{1}{j\omega C} = jX \tag{5.1}$$

$$C = -\frac{1}{\omega X} = -\frac{1}{2\pi \cdot f \cdot X} \tag{5.2}$$

$$C_{800\,\text{Hz}} = -\frac{1}{2\pi \cdot 800\,\text{Hz} \cdot (-47.240\,\Omega)} \approx 4.21\text{nF} \tag{5.3}$$

With an increasing frequency, the imaginary part turned towards positive values (see figures 5.30b and 5.31b).

The impedance of the system with a 58 cm lead was qualitatively the same, just shifted a little towards higher values for $\text{Re}(\underline{Z}) = R$ because of the elongated lead. The measured values for the 58 cm lead are visualized in figure 5.31.

Because the impedance analyzer was limited to a maximum frequency of 13 MHz, a fit to the values acquired for the tip electrode helped to extrapolate the input impedance up to a frequency of 42 or 64 MHz. Based on the results for the 45 cm lead, a linear fit was computed with Matlab. Taking into account the impedance values for all frequencies starting at 50 kHz, the computed expression was:

$$\text{Im}\{\underline{Z}(x)\} = 12.65 \cdot x - 414.8 \tag{5.4}$$

This equation described the connection between the imaginary part of the impedance ($f(x)$) and the real part (x). The fit and the impedance measurement data are shown

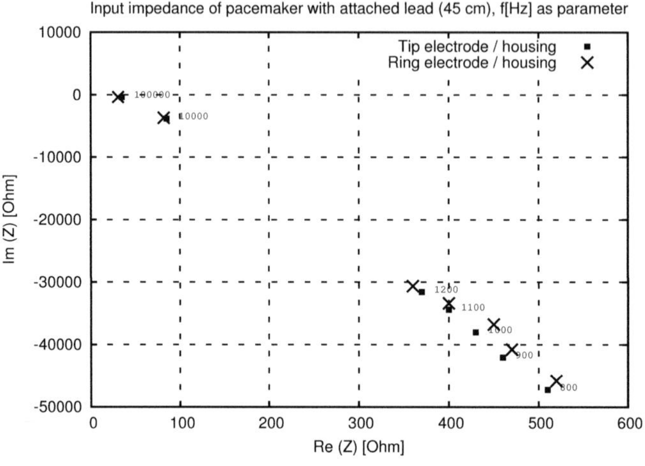

(a) Frequency range 800 Hz–100 kHz.

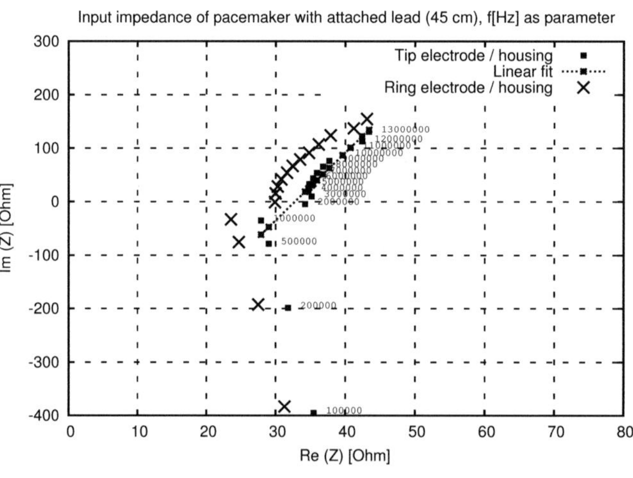

(b) Frequency range 100 kHz–13 MHz.

Figure 5.30.: Complex input impedance of St. Jude *Frontier II* connected to Medtronic *Capsurefix Novus* lead (45 cm).

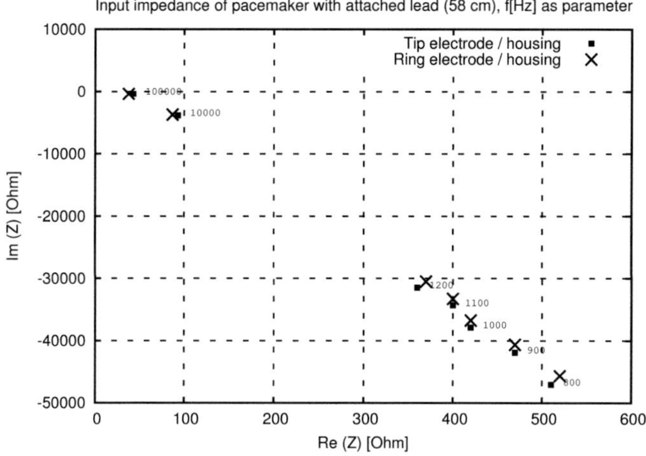

(a) Frequency range 800 Hz–100 kHz.

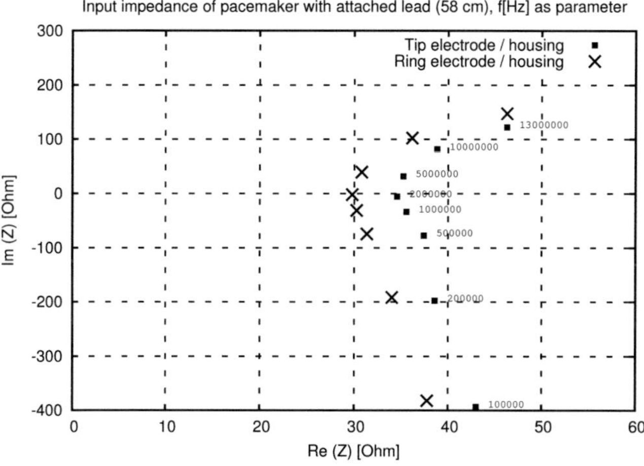

(b) Frequency range 100 kHz–13 MHz.

Figure 5.31.: Complex input impedance of St. Jude *Frontier II* connected to Medtronic *Capsurefix Novus* lead (58 cm).

in figure 5.30b as more detailed view for the frequency range above 100 kHz. Based on this model, the impedance of the system at 64 MHz was determined to $\underline{Z}_{64\,\mathrm{MHz}} = 87.72\,\Omega + j694\,\Omega$ and implemented in the computer model as series connection of a resistor and an inductor of $1.782\,\mu\mathrm{H}$.

Conclusion

The measurements of the input impedance of a commercial pacemaker lead system provided valuable information for designing realistic computer models. Although the used impedance analyzer did not cover the whole required frequency range, the acquired data could at least be used to extrapolate the necessary parameters.

In the low frequency region, the 5 nF value proposed by Osypka and Medtronic could be validated: at 800 Hz the impedance was about 4.21 nF. The frequency of cardiac pacing pulses is also about 1 kHz, which might explain, why the pacemaker producers communicated that figure. Furthermore, the findings highlight the importance of a realistic pacemaker modeling to achieve realistic results. As shown in the simulations, the lumped element approach is a suitable tool for this aspect when taking into account the frequency-dependence of the elements.

5.2.2. Bore Hole MRI

The first tranche of the experiments was carried out in the bore hole MRI device. In all experiments, a pacemaker was connected to the lead. The exact specifications of the used devices have been described in section 4.2.5 and 4.2.6. All descriptions regarding sides (left, right) imply the view of the patient.

Pacemaker with Lead

The experiment covered in this section was designed in the following way: the pacemaker unit had been placed on the right side of the carrier plate, the lead was coiled up one time and then oriented towards the left side of the phantom. In the vertical direction the whole arrangement was mounted in the middle of the phantom and covered by the same amount of saline above and below.

The observed temperature rise is shown in figure 5.32. After a small peak caused by the survey scan, the following T2-sequence provoked the first prominent rise of about $1\,°\mathrm{C}$. The next elevation corresponded to the 3D sequence and the longer lasting period

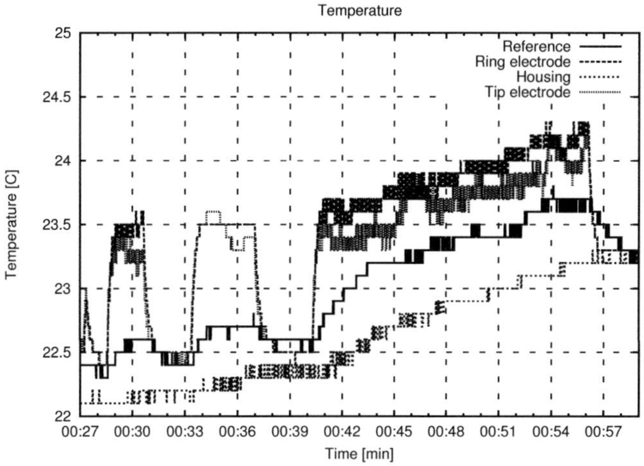

Figure 5.32.: Temperature around pacemaker with lead, inclination: $0°$, bore hole MRI.

could directly be correlated to the long T2-weighted sequence. The temperature at the ring electrode was for most of the time a little higher than at tip, which correlated with the simulation results of the complex lead model. The thermal changes at the housing and the reference point were less distinct, they mainly reflected the overall heating of the phantom filling. This heating was always caused in the bath by eddy currents - with or without a metal object present. In this experiment the probe next to the housing had an initial offset of $0.4\,°C$.

The next experiment was similar to the one described above but this time the carrier was inclined by an angle of $45°$. While the overall heating pattern stayed the same, the temperature rises at the electrodes were less steep and a constant difference between the tip and the ring electrode of $0.4\,°C$ could be observed (see figure 5.33). Again the housing and the reference probes only showed the general heating of the phantom filling.

The last measurement in the bore hole MRI presented here was performed with the carrier plate fully upright. The observed temperature changes were even less steep than in the $45°$ case but still present and again following the RF activation times (see figure 5.34).

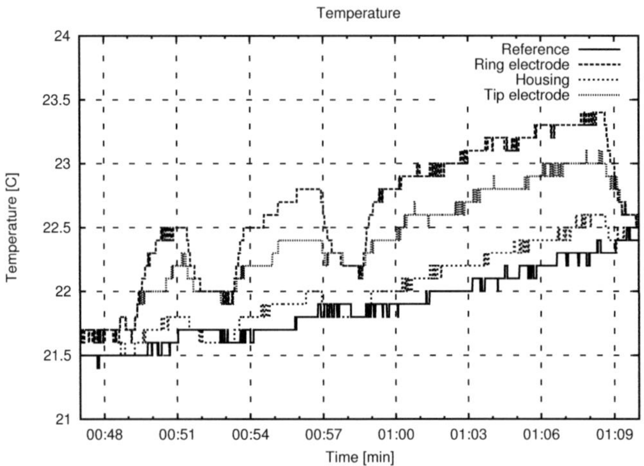

Figure 5.33.: Temperature around pacemaker with lead, inclination: $45°$, bore hole MRI.

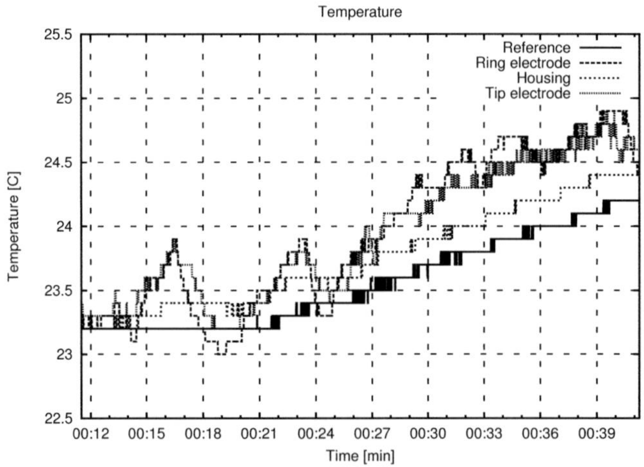

Figure 5.34.: Temperature around pacemaker with lead, inclination: $90°$, bore hole MRI.

Conclusion

The experiments in the bore hole MRI device supported and confirmed the results of the computer simulations. The highest and sharpest temperature gradients were found when the exposed objects were oriented flat in the x-y-plane and therefore perpendicular to the rotation plane of the B_1-fields. With an increased inclination of 45°, the increase became less prominent. The smallest heating was finally found in the total upright position.

When comparing the temperature at the beginning and at the end of each experiment one might think that even in the 90° case the tissue at the tips was heated up as much as in the first case. But the last sequence was unusually long and with its maximized SAR of 4 W/kg not typical for daily clinical practice. More relevant were the temperature curves in the first part of each session. In the 90° case, the gradients were much less distinct. For shorter sequences this implied a reduced risk for an accumulation of heat.

5.2.3. Open MRI

In the experiments presented now, the overall arrangement of the pacemaker and the lead was the same as in the bore hole MRI tests. The housing had been fixed at the right side and the lead, after one winding, pointed to the left side. For the first tests, the plate was mounted in a vertically centered position.

In the last two cases, the influence of the surrounding water layers should be evaluated. Because in a not-obese patient, the pacemaker unit could get placed close the body surface, the shielding effect of the covering fat layers would be significantly reduced. This was imitated by mounting the plate first close to the top surface and then at the very bottom of the phantom.

Pacemaker with Lead

The initial setup with the flat carrier plate produced nearly no heating (see figure 5.35). Additionally the MRI sequences did not induce any significant pattern in the slope. For the whole duration of the experiment, a total increase of 0.4 °C was recorded for all measurement locations.

After tilting the carrier for 45°, the response of the pacemaker-lead system did not change (see figure 5.36). Throughout the complete MRI protocol, only the overall temperature of the phantom changed. Like in the 0° case, a total heating of 0.4 °C could

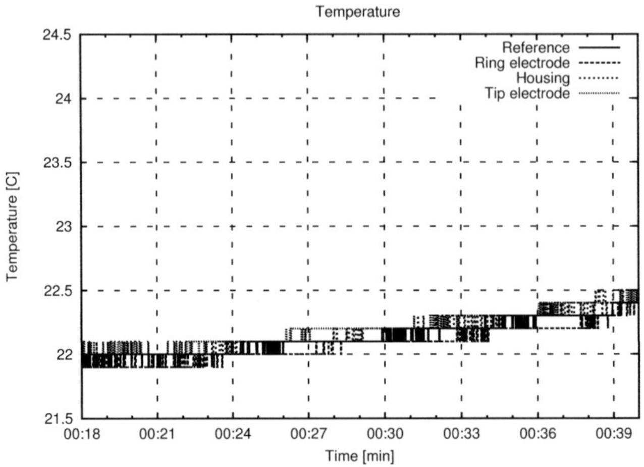

Figure 5.35.: Temperature around pacemaker with lead, inclination: $0°$, open MRI.

be found. When finally adjusting the plate to $90°$, the same effects as in the bore-hole MRI occurred (see figure 5.37). Heating was induced at the electrodes and it clearly correlated with the MRI sequences. Compared to the classic MRI, the increase of $0.5°C$ was nevertheless significantly lower in the open MRI system.

The last two experiments aimed at the influence of the over- and underlying liquid layers. For the configuration with the plate fixed close to the top surface, not remarkable changes were found (see figure 5.38). Just little temperature alterations were recorded at the pacemaker housing but they showed no real correlation to the RF activation periods. When mounted closed to the bottom, no remarkable changes besides a general heating of $0.1°C$ were observed (see figure 5.39).

Conclusion

The experiments in the open MRI device again confirmed the computer simulations. The temperature increase were less distinct than in the bore hole MRI: in the position that matched the regular bore hole MRI arrangement, the temperature just rose for about $0.4°C$. When the pacemaker/lead system was laid out flat, basically no heating occurred, only the overall phantom temperature changed, probably due to the waste heat of the device.

A significant influence of the over- and underlying water layers due to a variation of the

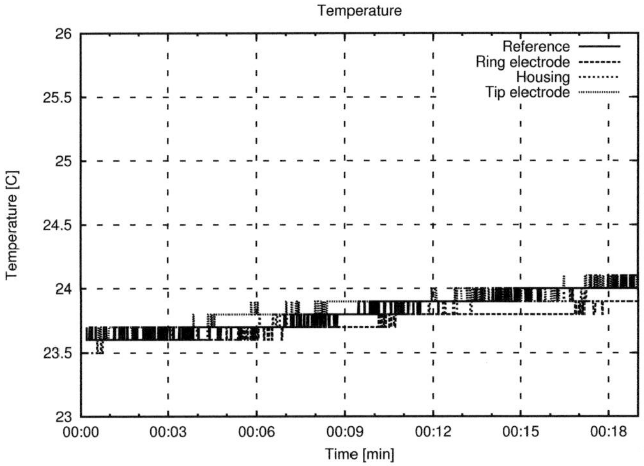

Figure 5.36.: Temperature around pacemaker with lead, inclination: 45°, open MRI.

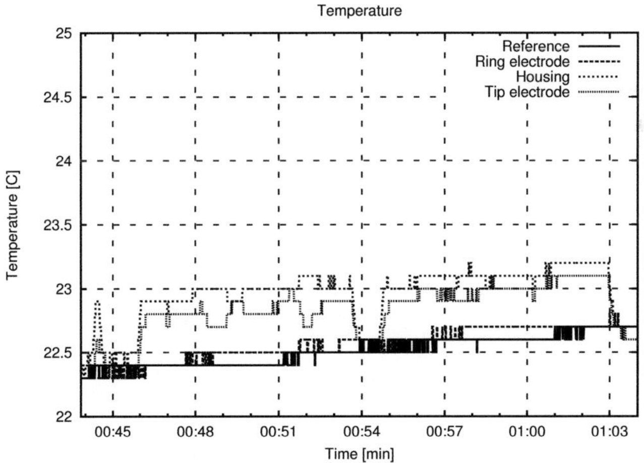

Figure 5.37.: Temperature around pacemaker with lead, inclination: 90°, open MRI.

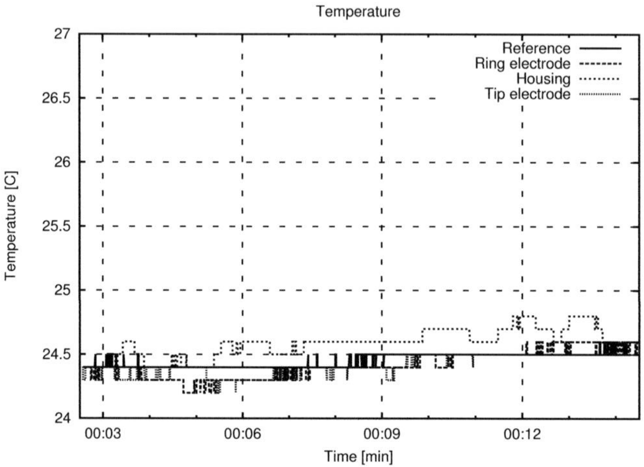

Figure 5.38.: Temperature around pacemaker with lead, inclination: $0°$, closer to top surface, open MRI.

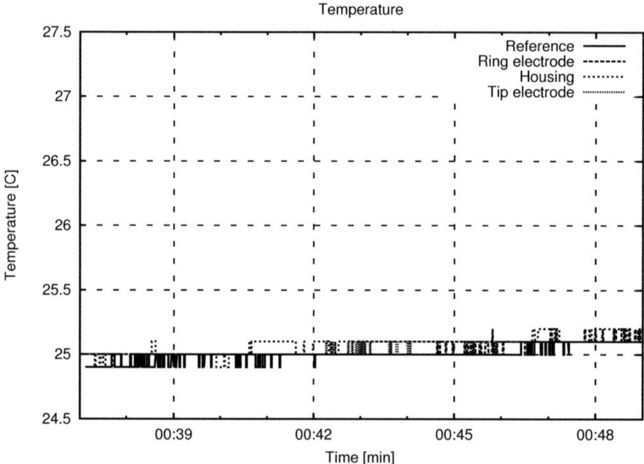

Figure 5.39.: Temperature around pacemaker with lead, inclination: $0°$, closer to bottom surface, open MRI.

vertical position could not be detected. Only the housing temperature changed in the position close to the bottom while the reason for this remained unclear since it did not correlate with the MRI sequences.

The experiments in the open MRI have shown that this type of MRI system causes considerably less heating in the vicinity of implants than the classic bore-hole device. Even with the high-SAR sequences employed here and with the pacemaker/lead oriented perpendicular to the rotating B_1-field, the heating found in the vicinity of the electrodes was far from values regarded as dangerous.

The comparison of these results with other studies is difficult. The only available publication on possible interactions of open MRI devices and cardiac pacemakers is the one of Luechinger et al. [52]. They had found higher temperature rises of more than $20°$ but used unusually high whole body SAR values of 1.31 W/kg. This was more than two times higher than the maximum approved clinical setting for the device. The work of Strach et al. in an low-field *Magnetom Concerto 0.2 T* (Siemens) only covered aspects like changes in lead impedance, capture threshold or battery voltage [53].

6. Summary

The aim of this work was the evaluation of heating in the surrounding of active cardiac implants during Magnetic Resonance Imaging. Repeatedly reported hazardous effects of heat induced by implants on the surrounding tissue in commonly used bore-hole MRI devices provoked a wide range of research efforts. As of today, the presence of such an implant still means a clear contraindication for MRI procedures. Due to the very high diagnostic value of MR imaging especially for patients with problems of the cardiac system, this is a very unfortunate situation.

Many aspects influence the genesis and amplitude of currents in the implants that later cause temperature elevations on galvanic contact points like the pacemakers housing or the electrodes. By means of computer simulation and in-vitro experiments, those aspects should be examined, observed patterns evaluated and possible workarounds identified.

Starting with a comprehensive literature survey and previous work here at the Institute of Biomedical Engineering, computer models for RF coils, phantoms, leads and pacemakers were developed. They allowed a detailed inspection of the occurring electromagnetic field distributions, the energy deposition and resulting temperature increases. The implant models were exposed to two types of RF fields. The first system mimicked a birdcage coil normally used as body coil in bore-hole MRI devices to generate the rotating B-Field orthogonal to the static B_0-field. The second was a newly developed representation of a RF coil for open MRI devices. Due to the different orientation of the rotating field with respect to the patient, the two coils allowed a systematic comparison of both MRI types and their interaction with implants. The objects under test were placed either in Plexiglas phantom models or a highly detailed anatomical voxel datasets, that both provided a dielectric milieu also prevailing in a human body.

The RF coils were stimulated with a continuous sinusoidal signal adjusted to 42 or 64 MHz in contrast to real MRI systems working with bursts of such signals. A comparable electromagnetic field strength inside the phantoms was achieved by scaling the continuous signal accordingly based on a comparative study with real MRI sequences.

The set of tested objects comprised several different types. Besides isolated straight wires, leads with helix shaped wires and detailed representations of the electrodes at the tip were evaluated. Furthermore, models of the pacemakers/lead system containing electronic components to reproduce the dielectric properties of those systems were developed. It became clear that due to the skin depth in the order of 0.1 mm the grid size near to the tip of the electrode had to be about 0.1mm leading to a significant increase in demands on storage capacity and calculation time. In addition averaging current densities and SAR over 10 g (corresponding to voxels of several millimeter side length) as recommended in the MRI guidelines is not adequate to describe the phenomena at the tip of the catheter properly. The resulting electromagnetic field and SAR distributions were significantly influenced by a several aspects. The most prominent impact had the orientation of the objects in relation to the RF fields. The higher the dB_1/dt along the course of the objects – e.g. at positions further from the center of the coils – the higher the observed induced currents and SAR values. For objects placed close to the body's center, the currents were substantially lower. Another property leading to lower SAR distributions was the encapsulation of the electrodes with fibrotic tissue. The reduced conductivity of this low-perfused tissue inhibited the transgression of induced currents into the cardiac tissue.

A complex lead/pacemaker model was based on a dissected and tested regular pacemaker model. The inclusion of electronic components allowed a realistic modeling of the input impedance of the lead/pacemaker assembly. The evaluation of a LR-low pass filter approach showed that with an appropriate adjustment of the filter's cut-off frequency the induced currents could be significantly reduced.

As already introduced with the description of its RF coil, a second MRI system was examined in this work: the open or panoramic MRI. Because its RF field is tilted by $90°$ compared to the classic bore hole devices, the pacemaker/lead arrangement inside a patient is no longer perpendicular to the rotation plane of the RF field but in parallel. A set of simulations with angles of $0°$, $45°$ and $90°$ was computed in both coils types. It affirmed the assumption of a significant reduction of induced currents for the open MRI system. The results of the computer simulations could be confirmed in a parallel in-vitro study together with the cardiology department of the University of Heidelberg. In this study, a pacemaker/lead combination was placed in a bore-hole and in an open MRI system – again at $0°$, $45°$ and $90°$ degrees. In the regular orientation of $0°$ in the open MRI device, basically no heating was found. This effect is further increased by a generally lower SAR caused by commonly used open MRI sequences.

When carrying out in-vitro experiments, the precise placement of the temperature probes turned out to be of particular importance. It was demonstrated that it is not the metal of the electrode that heats up during RF exposure but the tissue that is in direct contact to the electrodes. As observed in own measurements and confirmed in computer simulations, already a distance of 1 mm can lead to a serious underestimation of occurring temperature increases. The use of leads with incorporated thermocouple sensors directly bonded to the electrodes unfortunately proved to be impracticable. The exposed parts of the wires acted as antennas for the RF bursts themselves and the artifacts superimposed the microvolt signals induced by the temperature changes of a few degrees. As a consequence, the assessment of temperature changes in in-vivo experiments remains complicated.

In summary, the most effective measures against the induction of currents in implanted cardiac pacemaker devices proved to be the use of open MRI systems in combination with a pacemaker/lead system developed for MRI compatibility, for example, by including LR-low pass filters. The accumulation of deposited thermal energy can be effectively counteracted by even short breaks in the imaging procedures. Already a few seconds were enough for the heating to dissipate in the surrounding tissue. Because of the wide spectrum of possible implantation sites, lead paths and paccmaker/lead types, the classic bore hole MRI system seems inferior. The more advantageous RF field orientation and the generally lower SAR values of the open MRI system compensate a slightly reduced imaging quality. Important to highlight is that all acquired and presented results were achieved with MRI systems configured to maximum SAR settings and the models did not include any kind of perfusion. Therefore the configurations form worst case scenarios. In daily clinical practice, the used MRI sequences produce significantly lower SAR values and the duration of the procedures is shorter.

7. Outlook

The presented work provides a basis for future research in several fields:

The models for leads could be further improved. Now, the SAR distributions could only be determined for short parts of the detailed lead model. Newer generations of the computer hardware acceleration system might allow the computation of the helix shaped electrode following the whole course of a regular implantation procedure.

The yet simplified model of the pacemaker unit itself already provided valuable insights. A future project could examine, whether it is necessary to further improve the model and add more components to fully imitate a real-world device. Referring to the topic of realistic MRI sequences, a simplified pacemaker model might already be adequate.

A result similar to the one caused by the fibrotic tissue could be expected when the helix shaped electrode is mounted in the myocardium. The lower conductivity of the tissue compared to the blood in the heart's lumen could have the same effect. Together with the anyway higher SAR found around the ring electrode, these aspects could antagonize heating.

Another aspect only touched in a single in-vitro experiment yet is the topic of interventional MRI. As of today, the lead placement and extraction of biopsy samples is executed under X-ray guidance. The most negative side effect is the hazardous ionizing radiation, once for the patient and frequently for the physician. Another aspect is the achieved imaging information. X-ray only provides a projection and no 3D information. Additionally, MRI offers special sequences for ischemic and scar tissue, which allow highlighting of the regions of interest. Therefore the outcome of the procedures heavily relies on the experience of the physician. If a consecutive study would show, that with open MRI also interventional procedures are possible, not only the patient would avail but also the physician.

Despite the quality of modern computer simulations and initial in-vitro studies – for a final evaluation it still seems advisable to prove them in in-vivo experiments. Due to the highlighted issues with the temperature probe placement, a reliable real-time temperature measurement is not yet found. Perhaps an improved version of the Thermocouple system with a digitizing of the induced voltages as close to the tip as possible and later transfer via fiber optic cable could solve the electromagnetic compatibility problems and provide the neuralgic temperature values.

A. Abbreviations

AC	Alternating Current
API	Application Programming Interface
ASTM	American Society for Testing and Materials
CAD	Computer Aided Design
EM	Electromagnetic
FOV	Field of View
GPU	Graphics Processing Unit
HF	High Frequency
ICD	Implanted Cardioverter Defibrillator
MRI	Magnetic Resonance Imaging
PEC	Perfect Electric Conductor
RF	Radio Frequency
SAR	Specific Absorption Rate
SR	Saturation Recovery Sequence
TSE	Turbo Spin Echo Sequence
TSL	Tissue Simulating Liquid

B. Detailed company listing

Braun	B. Braun Melsungen AG, Melsungen, Germany
CST	CST Computer Simulation Technology, Darmstadt, Germany
FTDI	FTDI, Glasgow, United Kingdom
Geomagic	Geomagic, Research Triangle Park, NC, USA
Hewlett Packard	Hewlett Packard, Palo Alto, CA, USA
IT'IS Foundation	IT'IS Foundation (Zurich, Switzerland)
Mathworks	The Mathworks, Natick, MA, USA
Microsoft	Microsoft, Redmond, WA, USA
Medtronic	Medtronic, Minneapolis, MN, USA
Oil Center Research	Oil Center Research, Lafayette, LA, USA
OPTOcon	OPTOcon, Dresden, Germany
Osypka	Dr. Osypka GmbH, Rheinfelden-Herten, Germany
PCE	PCE Deutschland GmbH, Meschede, Germany
PeakTech	PeakTech, Ahrensburg, Germany
Siemens	Siemens Medical Solutions, Erlangen, Germany
St. Jude Medical	St. Jude Medical, Inc., St. Paul, MN, USA
Vitronic	Vitronic, Wiesbaden, Germany

C. List of Figures

D. List of Tables

E. Bibliography

[1] P. Davis, L. Crooks, M. Arakawa, R. McRee, L. Kaufman, and A. Margulis, "Potential hazards in NMR imaging: heating effects of changing magnetic fields and RF fields on small metallic implants," *AJR Am J Roentgenol*, vol. 137, pp. 857–860, 1981.

[2] M. Golombeck, O. Dössel, A. Staubert, and V. Tonnier, "Magnetic Resonance Imaging with Implanted Neurostimulators: A First Numerical Appoach Using Finite Integration Theory," in *Proc. International Symposium on Electromagnetic Compatibility 99*, 1999.

[3] E. Eriksen, M. Golombeck, S. Junge, and O. Dössel, "Simulation of a Birdcage and a Ceramic Cavity HF-resonator for high magnetic Fields in Magnetic Resonance Imaging," in *Biomedizinische Technik*, vol. 47-1, pp. 754–757, 2002.

[4] P. New, B. Rosen, T. Brady, F. Buonanno, J. Kistler, C. Burt, W. Hinshaw, J. Newhouse, G. Pohost, and J. Taveras, "Potential hazards and artifacts of ferromagnetic and nonferromagnetic surgical and dental materials and devices in nuclear magnetic resonance imaging," *Radiology*, vol. 147, pp. 139–148, 1983.

[5] i. s. ASTM, "ASTM F 2119 - 01: Standard test method for evaluation of mr image artifacts from passive implants," *ASTM International*, pp. 1207–1209, 2004.

[6] A. Klocke, J. Kemper, D. Schulze, G. Adam, and B. Kahl-Nieke, "Magnetic field interactions of orthodontic wires during magnetic resonance imaging (MRI) at 1.5 Tesla," *Journal of Orofacial Orthopedics = Fortschritte der Kieferorthopadie : Organ/Official Journal Deutsche Gesellschaft fur Kieferorthopadie*, vol. 66, pp. 279–287, 2005.

[7] M. Regier, J. Kemper, M. Kaul, M. Feddersen, G. Adam, B. Kahl-Nieke, and A. Klocke, "Radiofrequency-induced Heating near Fixed Orthodontic Appliances in High Field MRI Systems at 3.0 Tesla," *Journal of Orofacial Orthopedics/Fortschritte der Kieferorthopädie*, vol. 70, pp. 485–494, 2009.

[8] J. Starcukova, Z. Starcuk, H. Hubalkova, and I. Linetskiy, "Magnetic susceptibility and electrical conductivity of metallic dental materials and their impact on MR imaging artifacts," *Dental Materials : Official Publication of the Academy of Dental Materials*, vol. 24, pp. 715–723, 2008.

[9] Z. Cao, L. Chen, and X.-y. Gong, "[Artifacts from dental metal alloys in magnetic resonance imaging]," *Zhonghua yi xue za zhi*, vol. 88, pp. 1855–1858, 2008.

[10] W. Wang, B. Jiang, X. Wu, and J. J. Sun, "[Influences of three types of dental ceramic alloys on magnetic resonance imaging.]," *Zhongguo yi xue ke xue Yuan xue bao. Acta Academiae Medicinae Sinicae*, vol. 32, pp. 276–279, 2010.

[11] D. Destine, H. Mizutani, and Y. Igarashi, "Metallic artifacts in MRI caused by dental alloys and magnetic keeper," *Nihon Hotetsu Shika Gakkai Zasshi*, vol. 52, pp. 205–210, 2008.

[12] T. Taniyama, T. Sohmura, T. Etoh, M. Aoki, E. Sugiyama, and J. Takahashi, "Metal artifacts in MRI from non-magnetic dental alloy and its FEM analysis," *Dental Materials Journal*, vol. 29, pp. 297–302, 2010.

[13] s. i. ASTM, "ASTM F 2052 - 02: Standard test method for measurement of magnetically induced displacement force on medical devices in the magnetic resonance environment," *ASTM International*, pp. 1111–1116, 1993.

[14] M. Dujovny, N. Kossovsky, R. Kossowsky, R. Valdivia, J. Suk, F. Diaz, S. Berman, and W. Cleary, "Aneurysm clip motion during magnetic resonance imaging: in vivo experimental study with metallurgical factor analysis," *Neurosurgery*, vol. 17, pp. 543–548, 1985.

[15] R. Luechinger, F. Duru, M. Scheidegger, P. Boesiger, and R. Candinas, "Force and torque effects of a 1.5-Tesla MRI scanner on cardiac pacemakers and ICDs," *Pacing Clin Electrophysiol*, vol. 24, pp. 199–205, 2001.

[16] F. Shellock, J. Tkach, P. Ruggieri, and T. Masaryk, "Cardiac pacemakers, ICDs, and loop recorder: evaluation of translational attraction using conventional ("long-bore") and "short-bore" 1.5- and 3.0-Tesla MR systems," *J Cardiovasc Magn Reson*, vol. 5, pp. 387–397, 2003.

[17] R. Luechinger, F. Duru, V. Zeijlemaker, M. Scheidegger, P. Boesiger, and R. Candinas, "Pacemaker reed switch behavior in 0.5, 1.5, and 3.0 Tesla magnetic resonance imaging units: are reed switches always closed in strong magnetic fields?," *Pacing Clin Electrophysiol*, vol. 25, pp. 1419–1423, 2002.

[18] W. Irnich, B. Irnich, C. Bartsch, W. Stertmann, H. Gufler, and G. Weiler, "Do we need pacemakers resistant to magnetic resonance imaging?," *Europace*, vol. 7, pp. 353–365, 2005.

[19] M. Mollerus, G. Albin, M. Lipinski, and J. Lucca, "Ectopy in patients with permanent pacemakers and implantable cardioverter-defibrillators undergoing an MRI scan," *Pacing and Clinical Electrophysiology : PACE*, vol. 32, pp. 772–778, 2009.

[20] W. Irnich, "Re: Determinants of gradient fieldinduced current in a pacemaker lead system in a magnetic resonance imaging environment," *Heart Rhythm : the Official Journal of the Heart Rhythm Society*, vol. 5, pp. e2; author reply e2–3, 2008.

[21] H. Bassen and G. Mendoza, "In-vitro mapping of E-fields induced near pacemaker leads by simulated MR gradient fields," *Biomedical Engineering Online*, vol. 8, p. 39, 2009.

[22] P. Mansfield, "Multi-planar image formation using NMR spin echoes," *Journal of Physics C*, vol. 10, pp. L55–L58, 1977.

[23] S. Achenbach, W. Moshage, B. Diem, T. Bieberle, V. Schibgilla, and K. Bachmann, "Effects of magnetic resonance imaging on cardiac pacemakers and electrodes," *Am Heart J*, vol. 134, pp. 467–473, 1997.

[24] M. E. Ladd, H. H. Quick, P. Boesiger, and G. McKinnon, "RF Heating of Actively Visualized Catheters and Guidewires," in *Proceedings of the International Society for Magnetic Resonance in Medicine*, vol. S1, p. 474, 1998.

[25] J. Nyenhuis, A. Kildishev, J. Bourland, K. Foster, G. Graber, and T. Athey, "Heating near implanted medical devices by the MRI RF-magnetic field," in *Magnetics, IEEE Transactions on*, vol. 35, (Kyongju), pp. 4133–4135, 1999.

[26] P. Nordbeck and W. Bauer, "[Safety of cardiac pacemakers and implantable cardioverter-defibrillators in magnetic resonance imaging. Assessment of the aggregate function at 1.5 tesla]," *Deutsche Medizinische Wochenschrift (1946)*, vol. 133, pp. 624–628, 2008.

[27] P. Nordbeck, F. Fidler, I. Weiss, M. Warmuth, M. Friedrich, P. Ehses, W. Geistert, O. Ritter, P. Jakob, M. Ladd, H. Quick, and W. Bauer, "Spatial distribution of RF-induced E-fields and implant heating in MRI," *Magnetic Resonance in Medicine : Official Journal of the Society of Magnetic Resonance in Medicine / Society of Magnetic Resonance in Medicine*, vol. 60, pp. 312–319, 2008.

[28] J. Gimbel, "Magnetic resonance imaging of implantable cardiac rhythm devices at 3.0 tesla," *Pacing and Clinical Electrophysiology : PACE*, vol. 31, pp. 795–801, 2008.

[29] J. Stenschke, D. Li, M. Thomann, G. Schaefers, and W. Zylka, *Advances in Medical Engineering*, vol. 114, ch. A Numerical Investigation of RF Heating Effects on Implants During MRI Compared to Experimental Measurements, pp. 53–58. Springer Berlin Heidelberg, 2007.

[30] E. Mattei, M. Triventi, G. Calcagnini, F. Censi, W. Kainz, G. Mendoza, H. Bassen, and P. Bartolini, "Complexity of MRI induced heating on metallic leads: experimental measurements of 374 configurations," *Biomedical Engineering Online*, vol. 7, p. 11, 2008.

[31] E. Mattei, G. Calcagnini, F. Censi, M. Triventi, and P. Bartolini, "Methodological Issues on the Estimation of the MRI-Induced SAR in Tissues in Contact with Implanted Thin Metallic Structures," in *World Congress on Medical Physics and Biomedical Engineering, September 7 - 12, 2009, Munich, Germany*, IFMBE Proceedings, 2009.

[32] E. Neufeld, S. Kuhn, G. Szekely, and N. Kuster, "Measurement, simulation and uncertainty assessment of implant heating during MRI," *Physics in Medicine and Biology*, vol. 54, pp. 4151–4169, 2009.

[33] A. Kolandaivelu, M. Zviman, V. Castro, A. Lardo, R. Berger, and H. Halperin, "Non-invasive Assessment of Tissue Heating During Cardiac Radiofrequency Ablation Using MRI Thermography," *Circulation. Arrhythmia and Electrophysiology*, 2010.

[34] N. McDannold, "Quantitative MRI-based temperature mapping based on the proton resonant frequency shift: Review of validation studies," *International Journal of Hyperthermia*, vol. 21, pp. 533–546, 2005.

[35] W. Irnich, L. Batz, R. Muller, and R. Tobisch, "Electromagnetic interference of pacemakers by mobile phones," *Pacing Clin Electrophysiol*, vol. 19, pp. 1431–1446, 1996.

[36] I. Tandogan, B. Ozin, H. Bozbas, S. Turhan, R. Ozdemir, E. Yetkin, and E. Topal, "Effects of mobile telephones on the function of implantable cardioverter defibrillators," *Ann Noninvasive Electrocardiol*, vol. 10, pp. 409–413, 2005.

[37] I. Tandogan, A. Temizhan, E. Yetkin, Y. Guray, M. Ileri, E. Duru, and A. Sasmaz, "The effects of mobile phones on pacemaker function," *Int J Cardiol*, vol. 103, pp. 51–58, 2005.

[38] S. Seitz and O. Dössel, "Influence of body worn wireless mobile devices on implanted cardiac pacemakers," in *4th European Congress for Medical and Biomedical Engineering*, vol. 22, (Antwerp, Belgium), 2008.

[39] S. Seitz and O. Dössel, "Electromagnetic Fields near Implanted Cardiac Devices during Magnetic Resonance Imaging," in *IFMBE Proceedings World Congress on Medical Physics and Biomedical Engineering*, vol. 25/2, 2009.

[40] E. Adair, K. Mylacraine, and S. Allen, "Thermophysiological consequences of whole body resonant RF exposure (100 MHz) in human volunteers," *Bioelectromagnetics*, vol. 24, pp. 489–501, 2003.

[41] E. Adair and D. Black, "Thermoregulatory responses to RF energy absorption," *Bioelectromagnetics*, vol. Suppl 6, pp. 17–38, 2003.

[42] E. Adair, D. Blick, S. Allen, K. Mylacraine, J. Ziriax, and D. Scholl, "Thermophysiological responses of human volunteers to whole body RF exposure at 220 MHz," *Bioelectromagnetics*, vol. 26, pp. 448–461, 2005.

[43] W. Grill and J. Mortimer, "Electrical properties of implant encapsulation tissue," *Annals of Biomedical Engineering*, vol. 22, pp. 23–33, 1994.

[44] K. Stokes, J. Anderson, R. McVenes, and C. McClay, "The encapsulation of polyurethane-insulated transvenous cardiac pacemaker leads," *Cardiovascular Pathology*, vol. 4, pp. 163–171, 1995.

[45] R. Candinas, F. Duru, J. Schneider, T. Luscher, and K. Stokes, "Postmortem analysis of encapsulation around long-term ventricular endocardial pacing leads," *Mayo Clinic Proceedings. Mayo Clinic*, vol. 74, pp. 120–125, 1999.

[46] A. Jerzewski, P. Pattynama, P. Steendijk, J. Doornbos, A. de Roos, and J. Baan, "Development of an MRI-compatible catheter for pacing the heart: initial in vitro and in vivo results," *Journal of Magnetic Resonance Imaging : JMRI*, vol. 6, pp. 948–949, 1996.

[47] M. Ladd and H. Quick, "Reduction of resonant RF heating in intravascular catheters using coaxial chokes," *Magnetic Resonance in Medicine*, vol. 43, pp. 615–619, 2000.

[48] W. Greatbatch, V. Miller, and F. Shellock, "Magnetic resonance safety testing of a newly-developed fiber-optic cardiac pacing lead," *J Magn Reson Imaging*, vol. 16, pp. 97–103, 2002.

[49] J. Helfer, R. Gray, S. MacDonald, and T. Bibens, "Can pacemakers, neurostimulators, leads, or guide wires be MRI safe?: Technological concerns and possible resolutions," *Minim Invasive Ther Allied Technol*, vol. 15, pp. 114–120, 2006.

[50] B. Stevenson, W. Dabney, and C. Frysz, "Issues and design solutions associated with performing MRI scans on patients with active implantable medical devices," *Conference Proceedings : ... Annual International Conference of the IEEE Engineering in Medicine and Biology Society. IEEE Engineering in Medicine and Biology Society. Conference*, vol. 2007, pp. 6167–6170, 2007.

[51] H. Tandri, M. Zviman, S. Wedan, T. Lloyd, R. Berger, and H. Halperin, "Determinants of gradient field-induced current in a pacemaker lead system in a magnetic resonance imaging environment," *Heart Rhythm : the Official Journal of the Heart Rhythm Society*, vol. 5, pp. 462–468, 2008.

[52] R. Luechinger, V. A. Zeijlemaker, B. Mengiardi, and P. Boesiger, "RF heating of pacemaker leads: Do open MR systems perform better than close bore systems?," vol. 17, 2009.

[53] K. Strach, C. Naehle, A. Muhlsteffen, M. Hinz, A. Bernstein, D. Thomas, M. Linhart, C. Meyer, S. Bitaraf, H. Schild, and T. Sommer, "Low-field magnetic resonance imaging: increased safety for pacemaker patients?," *Europace : European Pacing, Arrhythmias, and Cardiac Electrophysiology : Journal of the Working Groups on Cardiac Pacing, Arrhythmias, and Cardiac Cellular Electrophysiology of the European Society of Cardiology*, 2010.

[54] M. F. Reiser, H. Hricak, and W. Semmler, "Magnetic Resonance Tomography," 2008.

[55] M. A. Bernstein and K. F. King, *Handbook of mri pulse sequences*. Elsevier Academic Press, 2004.

[56] O. Dössel, *Bildgebende Verfahren in der Medizin*. Berlin, Heidelberg, New York: Springer, 2000.

[57] T. Ibrahim, R. Lee, B. Baertlein, A. Kangarlu, and P. Robitaille, "Application of finite difference time domain method for the design of birdcage RF head coils using multi-port excitations," *Magnetic Resonance Imaging*, vol. 18, pp. 733–742, 2000.

[58] S. Smajic-Peimann and W. Zylka, *4th European Conference of the International Federation for Medical and Biological Engineering*, vol. 22, ch. Co-simulation approach for the design of MRI RF coils and its application to local SAR distribution, pp. 2636–2639. Springer Berlin Heidelberg, 2009.

[59] S. Lee, M. Cho, C. Moon, and H. Park, "A convex gradient coil design for vertical field open MRI," in *Engineering in Medicine and Biology Society, 2000. Proceedings of the 22nd Annual International Conference of the IEEE*, vol. 3, pp. 2161–2162, 2000.

[60] M. Zhu, L. Xia, F. Liu, and S. Crozier, "Deformation-Space Method for the Design of Biplanar Transverse Gradient Coils in Open MRI Systems," *Magnetics, IEEE Transactions on*, vol. 44, pp. 2035–2041, 2008.

[61] H. Fujita, W. O. Braum, and M. A. Morich, "Novel quadrature birdcage coil for a vertical B0 field open MRI system," *Magnetic Resonance in Medicine*, vol. 44, pp. 633–640, 2000.

[62] B. McCarten, J. Carlson, J. Fehn, M. Arakawa, and L. Kaufman, "Open design and flat cross sectional RF transmit coil for transverse magnet based MRI systems," in *Nuclear Science Symposium and Medical Imaging Conference, 1993., 1993 IEEE Conference Record.*, pp. 1706–1707, 1993.

[63] E. B. Boskamp, "Flat RF body coil design for open MRI," in *Engineering in Medicine and Biology Society, 2000. Proceedings of the 22nd Annual International Conference of the IEEE*, vol. 3, pp. 2387–2389, 2000.

[64] G. L. Khym, H. J. Yang, and C. H. Oh, "The design of a body RF coil for low-field open MRI using pseudo electric dipole radiation and simulated annealing," *Current Applied Physics*, vol. In Press, Corrected Proof, 2010.

[65] J. C. Maxwell, "A Dynamical Theory of the Electromagnetic Field," *Philosophical Transactions of the Royal Society of London*, vol. 155, pp. 459–512, 1865.

[66] M. Golombeck, *Feldtheoretische Studien zur Patientensicherheit bei der Magnetresonanztomographie und der Elektrochirugie.* PhD thesis, 2003.

[67] A. Schwab and F. Imo, *Begriffswelt der Feldtheorie.* Berlin, Heidelberg: Springer, 2002.

[68] SPEAG (Schmid & Partner Engineering AG), *SEMCAD X Reference Manual (Ver. 14.2).* Zürich, 2010.

[69] H. Pohl, "Giant polarization in high polymers," *Journal of Electronic Materials*, vol. 15, pp. 201–203, 1986.

[70] N. Ida and J. P. A. Bastos, *Electro-Magnetics and Calculation of Fields.* Springer, 1997.

[71] C. Younkin, "Multiphase MP35N alloy for medical implants," *Journal of Biomedical Materials Research*, vol. 8, pp. 219–226, 1974.

[72] K. Yee, "Numerical solution of initial boundary value problems involving Maxwell's equations in isotropic media," *Antennas and Propagation, IEEE Transactions on*, vol. 14, pp. 302–307, 1966.

[73] C. Buccella, V. De Santis, and M. Feliziani, "Prediction of Temperature Increase in Human Eyes Due to RF Sources," *IEEE Transactions on Electromagnetic Compatibility*, vol. 49, pp. 825–833, 2007.

[74] M. Busch, W. Vollmann, T. Bertsch, R. Wetzler, A. Bornstedt, B. Schnackenburg, J. Schnorr, D. Kivelitz, M. Taupitz, and D. Gronemeyer, "On the heating of inductively coupled resonators (stents) during MRI examinations," *Magn Reson Med*, vol. 54, pp. 775–782, 2005.

[75] V. Flyckt, B. Raaymakers, H. Kroeze, and J. Lagendijk, "Calculation of SAR and temperature rise in a high-resolution vascularized model of the human eye and orbit when exposed to a dipole antenna at 900, 1500 and 1800 MHz," *Phys Med Biol*, vol. 52, pp. 2691–2701, 2007.

[76] V. Flyckt, B. Raaymakers, and J. Lagendijk, "Modelling the impact of blood flow on the temperature distribution in the human eye and the orbit: fixed heat transfer coefficients versus the Pennes bioheat model versus discrete blood vessels," *Phys Med Biol*, vol. 51, pp. 5007–5021, 2006.

[77] DIN, "DIN EN 60601-2-33 - Medizinische elektrische Geräte: Besondere Festlegung für die Sicherheit von Magnetresonanzgeräten für die medizinische Diagnostik," *Deutsches Instituit für Normung e. v. und VDE Verband für Elektrotechnik Elektronik Informationstechnik e. V.*, vol. DIN EN 60601-2-33 + Entwurf, 2003.

[78] DIN, "DIN EN 62209-1 Sicherheit von Personen in hochfrequenten Feldern von handgehaltenen und am Körper getragenen schnurlosen Kommunikationsgeräten - Körpermodelle, Messgeräte und Verfahren," vol. Teil 1: Verfahren zur Bestimmung der spezifischen, 2006.

[79] S. Silbernagl and A. Despopoulos, *Taschenatlas Physiologie*. Thieme, 2007.

[80] B. Harmon, A. Corder, R. Collins, G. Gobe, J. Allen, D. Allan, and J. Kerr, "Cell death induced in a murine mastocytoma by 42-47 degrees C heating in vitro: evidence that the form of death changes from apoptosis to necrosis above a critical heat load," *International Journal of Radiation Biology*, vol. 58, pp. 845–858, 1990.

[81] J. Roti Roti, "Cellular responses to hyperthermia (40-46 degrees C): cell killing and molecular events," *International Journal of Hyperthermia : the Official Journal of European Society for Hyperthermic Oncology, North American Hyperthermia Group*, vol. 24, pp. 3–15, 2008.

[82] D. Koncan, J. Rifel, G. Drevensek, S. Kocijancic, S. Ogorelec, and M. V. Budihna, "Thermal conductivity of the porcine heart tissue," *Pflügers Archiv European Journal of Physiology*, vol. 440, pp. r143–r144, 2000.

[83] H. Pennes, "Analysis of tissue and arterial blood temperatures in the resting human forearm," *Journal of Applied Physiology*, vol. 1, pp. 93–122, 1948.

[84] J. Beuthan and O. Minet, "Phenomenological statistics of laser irradiation related metabolic changes in guinea pig livers," *Biomedizinische Technik. Biomedical Engineering*, vol. 49, pp. 238–241, 2004.

[85] L. F. Romero, O. Trelles, and M. A. Trelles, "Real-Time Simulation for Laser-Tissue Interaction Model," in *Parallel Computing: Current & Future Issues of High-End Computing, Proceedings of the International Conference ParCo 2005, 13-16 September 2005, Department of Computer Architecture, University of Malaga, Spain*, vol. 33, pp. 415–422, 2005.

[86] Z. Ji-zhuang and Z. Xue-xue, "Dynamic Modeling of Tissue Ablation with Continuous Wave CO2 Laser," in *Bioinformatics and Biomedical Engineering, 2007. ICBBE 2007. The 1st International Conference on*, pp. 1057–1060, 2007.

[87] L. P. Walsh, J. K. Anderson, M. R. Baker, B. Han, J.-T. Hsieh, Y. Lotan, and J. A. Cadeddu, "In Vitro Assessment of the Efficacy of Thermal Therapy in Human Renal Cell Carcinoma," *Urology*, vol. 70, pp. 380–384, 2007.

[88] M. Pop, S. Davidson, M. Gertner, M. Jewett, M. Sherar, and M. Kolios, *Biomedical Simulation*, ch. A Theoretical Model for RF Ablation of Kidney Tissue and Its Experimental Validation, pp. 119–129. 2010.

[89] I. Chang, "Considerations for thermal injury analysis for RF ablation devices," *The Open Biomedical Engineering Journal*, vol. 4, pp. 3–12, 2010.

[90] H. Shafiee, P. A. Garcia, and R. V. Davalos, "A Preliminary Study to Delineate Irreversible Electroporation From Thermal Damage Using the Arrhenius Equation," *Journal of Biomechanical Engineering*, vol. 131, p. 074509, 2009.

[91] R. C. Lee, E. G. Cravalho, and J. F. Burke, *Electrical Trauma: the Pathophysiology, Manifestations and Clinical Management*. Cambridge University Press, 1992.

[92] R. Agah, J. Pearce, A. Welch, and M. Motamedi, "Rate process model for arterial tissue thermal damage: implications on vessel photocoagulation," *Lasers in Surgery and Medicine*, vol. 15, pp. 176–184, 1994.

[93] C. Gabriel, S. Gabriel, and E. Corthout, "The dielectric properties of biological tissues: I. Literature survey," *Physics in Medicine and Biology*, vol. 41, pp. 2231–2250, 1996.

[94] S. Gabriel, R. Lau, and C. Gabriel, "The dielectric properties of biological tissues: II. Measurements in the frequency range 10 Hz to 20 GHz," *Physics in Medicine and Biology*, vol. 41, pp. 2251–2269, 1996.

[95] S. Gabriel, R. Lau, and C. Gabriel, "The dielectric properties of biological tissues: III. Parametric models for the dielectric spectrum of tissues," *Physics in Medicine and Biology*, vol. 41, pp. 2271–2293, 1996.

[96] J. Woloszko, K. Stalder, and I. Brown, "Plasma characteristics of repetitively-pulsed electrical discharges in saline solutions used for surgical procedures," *Plasma Science, IEEE Transactions on*, vol. 30, pp. 1376–1383, 2002.

[97] H. Bassen, W. Kainz, G. Mendoza, and T. Kellom, "MRI-induced heating of selected thin wire metallic implants? laboratory and computational studies? findings and new questions raised," *Minim Invasive Ther Allied Technol*, vol. 15, pp. 76–84, 2006.

[98] A. International, *Materials and Coatings for Medical Devices: Cardiovascular*. 2009.

[99] E. Hecht, *Optik*. München: Oldenbourg, 2009.

[100] J. Craig, P. Simmons, S. Patel, and A. Tomlinson, "Refractive Index and Osmolality of Human Tears," *Optometry & Vision Science*, vol. 72, pp. 718–724, 1995.

[101] A. Bolz and W. Urbaszek, *Technik in der Kardiologie: eine interdisziplinäre Darstellung*. Berlin, Heidelberg: Springer, 2002.

[102] P. Nordbeck, I. Weiss, P. Ehses, O. Ritter, M. Warmuth, F. Fidler, V. Herold, P. Jakob, M. Ladd, H. Quick, and W. Bauer, "Measuring RF-induced currents inside implants: Impact of device configuration on MRI safety of cardiac pacemaker leads," *Magnetic Resonance in Medicine : Official Journal of the Society of Magnetic Resonance in Medicine / Society of Magnetic Resonance in Medicine*, vol. 61, pp. 570–578, 2009.

[103] M. Ackerman, V. M. Spitzer, A. L. Scherzinger, and D. G. Whitlock, "The Visible Human data set: an image resource for anatomical visualization," *Medinfo*, vol. 8, pp. 1195–1198, 1995.

[104] S. Bauer, D. Keller, F. M. Weber, P. Tri Dung, G. Seemann, and O. Dössel, "How do tissue conductivities impact on forward-calculated ECGs? An efficient prediction based on principal component analysis," in *IFMBE Proceedings World Congress on Medical Physics and Biomedical Engineering*, vol. 25/4, pp. 641–644, 2009.

[105] D. U. J. Keller, F. M. Weber, G. Seemann, and O. Dössel, "Ranking the Influence of Tissue Conductivities on Forward-Calculated ECGs," *IEEE Transactions on Biomedical Engineering*, vol. 57, pp. 1568–1576, 2010.

[106] M. Cristy and Oak Ridge National Laboratory, "Mathematical Phantoms Representing Children of Various Ages for Use in Estimates of Internal Dose," tech. rep., 1980.

[107] K. Eckerman, M. Cristy, and J. Ryman, "The ORNL Mathematical Phantom Series," tech. rep., 1996.

[108] E. Mattei, M. Triventi, G. Calcagnini, F. Censi, W. Kainz, H. Bassen, and P. Bartolini, "Temperature and SAR measurement errors in the evaluation of metallic linear structures heating during MRI using fluoroptic probes," *Physics in Medicine and Biology*, vol. 52, pp. 1633–1646, 2007.

[109] i. s. ASTM, "ASTM F 2182 - 02a: Standard test method for radio frequency induced heating near passive implants during magnetic resonance imaging," *ASTM International*, pp. 1275–1282, 1993.

[110] M. Konings, L. Bartels, H. Smits, and C. Bakker, "Heating around intravascular guidewires by resonating RF waves," *J Magn Reson Imaging*, vol. 12, pp. 79–85, 2000.

[111] C.-K. Chou, G.-W. Chen, A. Guy, and K. Luk, "Formulas for preparing phantom muscle tissue at various radiofrequencies," *Bioelectromagnetics*, vol. 5, pp. 435–441, 1984.

[112] FDA U.S. Food and Drug Administration, "Device Approvals and Clearances - St. Jude Frontier(TM) Biventricular Cardiac Pacing System including the Frontier(TM) Model 5508 and 5508L Pulse Generators, the Aescula(TM) LV Model 1055K Lead and the Model 3307 v4.4.2m programmer software for use with the Model 3500/3510 Programmer - P030035," 2004.

[113] M. Grafmüller, S. Seitz, and O. Dössel, "Adaption of Generic Anatomic Organ Models on Patient Specific Data Sets," in *IFMBE Proceedings World Congress on Medical Physics and Biomedical Engineering*, vol. 25/4, 2009.

[114] M. Grafmüller, "Generierung realistischer Körpermodelle für die numerische Feldberechnung," Master's thesis, 2009.

[115] S. Seitz and O. Dössel, "Numerical Modeling of Current Distribution in and near the Tips of Cardiac Pacemaker Electrodes during Magnetic Resonance Imaging," in *Proc. Computers in Cardiology*, 2009.

[116] M. Fütterer, "Einfluss der Hochfrequenz-Anregungsmuster auf die Erwärmung an Herzschrittmacherelektroden während der Magnetresonanztomographie," Master's thesis, 2010.

[117] M. Fütterer and S. Seitz, "Influence of RF-excitation patterns during Magnetic Resonance Imaging on heating at the tip of pacemaker electrodes," in *Gemeinsame Jahrestagung der Deutschen, der Österreichischen und der Schweizerischen Gesellschaft für Biomedizinische Technik*, 2010.